AF802912

Tourism and Foreign Direct Investment

This is the first volume to examine and shed significant light on the issues, challenges and prospects presented by foreign direct investment (FDI) in tourism – a topic of increasing relevance in light of the COVID-19 pandemic and economic consequences.

This book addresses the need to examine the TFDI phenomenon considering resilient tourism development approaches and includes discussion on impacts of a rapidly evolving global pandemic on tourism. Recognising that its impact on tourism has only just begun, the book includes early estimates of the damage to the tourism economy and TFDI over 2020 and beyond. It considers how the COVID-19 pandemic may change society, the economy, and tourism, and how some of the key research needs to understand these changes and contribute to a more sustainable post-pandemic tourism sector. Through a collection of chapters focusing on principles, practices and a diverse range of destination case studies, the book considers TFDI from the economic, social and environmental and regulatory perspectives. Thus, the book will advance understanding of the positive and negative impacts of TFDI as well as how emerging problems and wider implications for local economies and communities can be managed more sustainably, while also discussing the transformative opportunities offered by TFDI regarding key economic, social and environmental issues.

This will be essential reading for upper-level students and researchers in the field of tourism development.

H. Cristina Jönsson is a Lecturer at The University of the West Indies, Barbados. She holds a PhD in Tourism awarded by the Sheffield Business School at Sheffield Hallam University, UK. Before joining The University of the West Indies, Cristina was a municipal tourism manager in Sweden. She has extensive knowledge and experience from planning, development and management in the international tourism sector and has worked with both private and public sector organisations in various countries. Cristina's wide cultural familiarity along with mastering seven languages has been a great asset in all projects, research and collaborations, as well as translation and interpretation work. Her current research interests include foreign direct investment (FDI) in tourism, small island tourism, sport tourism, post pandemic migration and tourism mobilities.

Routledge Advances in Tourism
Edited by Stephen Page, Hertfordshire Business School, University of Hertfordshire, UK

Lifelong Learning for Tourism
Concepts, Policy and Implementation
Edited by Violet V. Cuffy, David Airey and Georgios C. Papageorgiou

Hosting the Olympic Games
Uncertainty, Debates, and Controversy
Edited by Marie Delaplace and Pierre-Olaf Schut

Encountering Nazi Tourism Sites
Derek Dalton

Evaluating the Local Impacts of the Rio Olympics
Edited by Marcelo Neri

Mountain Resort Marketing and Management
Armelle Solelhac

Ageing and the Visitor Economy
Global Challenges and Opportunities
Stephen J. Page and Joanne Connell

Asian Mobilities Consumption in a Changing Arctic
Edited by Young-Sook Lee

Tourism and Foreign Direct Investment
Issues, Challenges and Prospects
Edited by H. Cristina Jönsson

For more information about this series, please visit: www.routledge.com/advances-in-tourism/book-series/SE0538

Tourism and Foreign Direct Investment

Issues, Challenges and Prospects

Edited by
H. Cristina Jönsson

Routledge
Taylor & Francis Group
LONDON AND NEW YORK

First published 2022
by Routledge
2 Park Square, Milton Park, Abingdon, Oxon OX14 4RN

and by Routledge
605 Third Avenue, New York, NY 10158

Routledge is an imprint of the Taylor & Francis Group, an informa business

© 2022 selection and editorial matter, H. Cristina Jönsson;
individual chapters, the contributors

The right of H. Cristina Jönsson to be identified as the author of the editorial material, and of the authors for their individual chapters, has been asserted in accordance with sections 77 and 78 of the Copyright, Designs and Patents Act 1988.

All rights reserved. No part of this book may be reprinted or reproduced or utilised in any form or by any electronic, mechanical, or other means, now known or hereafter invented, including photocopying and recording, or in any information storage or retrieval system, without permission in writing from the publishers.

Trademark notice: Product or corporate names may be trademarks or registered trademarks, and are used only for identification and explanation without intent to infringe.

British Library Cataloguing-in-Publication Data
A catalogue record for this book is available from the British Library

Library of Congress Cataloging-in-Publication Data
A catalog record has been requested for this book

ISBN: 978-0-367-72594-5 (hbk)
ISBN: 978-0-367-72601-0 (pbk)
ISBN: 978-1-003-15549-2 (ebk)

DOI: 10.4324/9781003155492

Typeset in Times New Roman
by codeMantra

Contents

Contributors

Dr James Antwi is a Lecturer of Economics at the Ghana Institute of Management and Public Administration (GIMPA). James has over 24 years of consulting, industry, and teaching experience. Before joining GIMPA, he worked for the State of Ohio Treasury for 11 years. He is the CEO of POMI Consultancy Services Ltd. James has a Ph.D. in Economics and a Masters in Economics and Applied Econometrics from the University of Delaware, a Masters in Financial Economics from Ohio University, an MBA in Administration from Franklin University, a Masters in System Science from the University of Ottawa and a Bachelors in Mathematical Science and Applied Statistics from the University of Toronto. James is a project management professional (PMP) certified.

Dr Susana Cró is an Assistant Professor at School of Social Sciences in the University of Évora, Portugal. She holds a PhD in Tourism awarded by the Institute of Geography and Spatial Planning (IGOT) – University of Lisbon, Portugal. She is an integrated investigator in CEFAGE – Centre for Advanced Studies in Management and Economics of the University of Évora. Her current research activities include crisis and disasters management, security and safety in tourism industry, performance measurement and management in hospitality and tourism, tourism promotion and foreign direct investment (FDI).

Professor Larry Dwyer publishes widely in the areas of tourism economics, management, policy and planning. Larry is a Fellow and past President of the International Academy for Study of Tourism. He is past President of the International Association for Tourism Economics, and currently serves on its International Advisory Board.

Dr Sheereen Fauzel is a Senior Lecturer at the University of Mauritius. Having completed a PhD in International Economics from the University of Mauritius, a Masters degree in Banking and Finance and a Bachelor degree in Economics and Finance, her areas of expertise are International Economics, Development Economics, Tourism Economics and related areas. Sheereen has participated in international conferences and has publications in notable international journals of Business, Economics and Tourism. She is a reviewer for a number of refereed journals including *Current Issues in Tourism* and *Journal of Hospitality Marketing and Management*, among others.

Dr Maneka Jayasinghe is a Lecturer in Economics in the Asia Pacific Collage of Business and Law at Charles Darwin University, Australia. Her research interests include consumer demand analysis, tourism analysis, poverty and inequality and gender. She has published a number of papers in international refereed journals such as *Social Indicators Research*, *Applied Economics* and *Food Policy*. She has also received a number of research grants and awards including best paper and best presenter awards.

Dr H. Cristina Jönsson is a Lecturer at The University of the West Indies, Barbados. She holds a PhD in Tourism awarded by the Sheffield Business School at Sheffield Hallam University, UK. Before joining The University of the West Indies, Cristina was a municipal tourism manager in Sweden. She has extensive knowledge and experience from planning, development and management in the international tourism sector and has worked with both private and public sector organisations in various countries. Cristina's wide cultural familiarity along with mastering seven languages has been a great asset in all projects, research and collaborations, as well as translation and interpretation work. Her current research interests include foreign direct investment (FDI) in tourism, small island tourism, sport tourism, post pandemic migration and tourism mobilities.

Dr António Miguel Valente Martins is an Assistant Professor at the Faculty of Social Sciences in the University of Madeira (Portugal), where he coordinates the management and economics department. He holds a PhD in Management awarded by the University of Porto, Portugal. António is an integrated investigator in CEEAplA – Atlantic Applied Economic Studies Center. His current research interests include crisis and disasters management, security and safety in tourism industry, performance measurement and management in hospitality and tourism, finance, real estate management and foreign direct investment (FDI).

Dr Bryna Meivitawanli completed a double degree bachelor program from University of Indonesia and University of Groningen, the Netherlands. She was awarded a PhD in management at the age of 25 from Wuhan University of Technology, China. She received full scholarship from the Chinese Government for her doctoral degree. Bryna joined Binus Business School, an AACSB accredited business school, in 2020 as subject content coordinator and lecturer in the International Undergraduate program. Her research interests include marketing, economics and tourism. Most of her publications are focused on foreign direct investment.

Dr Maja Nikšić Radić is an Associate Professor currently working at the Faculty of Tourism and Hospitality Management in Opatija, University of Rijeka in Croatia under the Department of International Economics and Development. She holds a PhD in international economics, which is her main area of interest. Over the years, Maja has been professionally improving in fields like tourism FDI, FDI and economic growth, FDI and sustainable development and global economic challenges. She has published numerous scientific articles and participated in numerous scientific conferences. Maja's primary research areas are FDI in tourism, terrorism and migration.

Dr Nicola J. Palmer is an academic based in the Doctoral School at Sheffield Hallam University and aligned to the Culture and Creativity Research Institute. Nicola has research interests in tourism governance – the application of power, resources, rules and coordination and cooperation among numerous tourism actors. Specifically, her published work draws on political economy and political ecology to generate new knowledge on ways in which social status, legitimacy and power play a key role in the dynamics of sustainable community-based tourism development by stimulating local economies, conserving the environment, developing people's livelihoods and changing lives.

Professor Eliyathamby A Selvanathan is currently the Director of the Economics Policy Analysis Program (EPAP) at the Griffith Business School and Professor of Econometrics at Griffith University. He is one of the authors of the best-selling textbook *Business Statistics* published by Cengage. His research interests include applied econometrics, time series modelling, tourism economics, consumer demand models and stochastic index numbers. He has published eight research monographs and well over 100 articles in leading journals such as *Review of Economic Studies, Journal of Econometrics, Journal of Economics and Business Statistics, Review of Economic Statistics, Tourism Analysis* and *Tourism Economics.*

Professor Saroja Selvanathan is a Professor of Econometrics in the Economics and Business Statistics Discipline at Griffith University. She is also the Co-Editor-in-Chief of the *Journal of the Asia Pacific Economy.* Her research interests include consumer demand, time series analysis, tourism analysis and applied econometrics. Professor Selvanathan has published five research monographs and over 75 journal articles in the broad area of applied econometrics. She is a co-author of the Australian best-selling textbook, *Business Statistics*, published by Cengage. Her publications have appeared in internationally referred journals such as *Review of Economic Statistics, Empirical Economics, Economic Modelling, Tourism Analysis* and *Tourism Economics.*

Sara Tasneem has 12 years' experience as an Assistant Professor in Entrepreneurial Economics at Dhaka School of Economics in Bangladesh. Sara completed her BSS and MSS in Economics from the University of Dhaka in 2002 and 2003, respectively. She received her MBA in Finance from Daffodil International University in 2011. At present she is pursuing PhD research in the Department of Banking and Insurance, University of Dhaka. She believes in dedication and honesty while working. Tasneem is a powerful force in the workplace and uses her positive attitude and tireless energy to encourage others to work hard and succeed. Her professional goal is to be academically developed. She enjoys watching news and analysing socio-economic phenomena. She is inspired daily by her husband and their son.

Acknowledgements

The world is a better place thanks to people who want to share their experiences and knowledge. What makes it even better are people who share the gift of their time, guidance, collegiality and mentorship.

This book is the result of true teamwork across disciplines. I am deeply grateful to the contributors from around the world, who accepted my invitation to write a chapter for this book on *Tourism Foreign Direct Investment.* You made this book possible.

To Professor Philmore Alleyene, a true colleague and mentor. When we first met, he saw a junior academic eager to learn and eager to grow. He always pushed me to engage in research and encouraged me to write this book. Thank you for the support and for being an inspiration.

I have to thank my dear friend, colleague and mentor Professor Lee Jolliffe. From encouraging me to embark on this project, giving me advice from its inception to guiding me through the book editing process. She was as important to this book materialising as I was.

For all this I am truly grateful.

H. Cristina Jönsson
Bridgetown, Barbados

Part I

Introduction

1 An Overview of Tourism-Related Foreign Direct Investment (TFDI)

H. Cristina Jönsson

Conceptualising and Explaining Tourism-Related Foreign Direct Investment

While the principle of FDI is a commonly used term, definitions of FDI have a variety of interpretations. According to the Organisation for Economic Co-operation and Development (OECD, 1996), there is consistency in definitions of FDI used by itself and international development organisations, the United Nations Council for Tourism and Development (UNCTAD) and the International Monetary Fund (IMF). UNCTAD (2006) explains FDI as the objective of capital invested for the purpose of acquiring a lasting interest and control by a resident entity in one economy or an enterprise resident in an economy other than that of the investor. In this long-term relationship between the investor and the enterprise, the crucial aspects distinguishing FDI from other types of investment are location and control. A more succinct definition of FDI is provided by Economy Watch (2010), "the injection of foreign funds into an enterprise that operates in a different country of origin from the investor"; yet the issue of direct control should not be overlooked.

FDI is the inflow of capital into and out of a particular sector of a country and signifies direct control of an enterprise by investors. This type of investment comprises (a) settler-type investment where the investor and their capital move to a country, (b) "putting-out" investment where ownership and control resides abroad but without links to other enterprises, and (c) the multinational corporation as an extension of the corporation into foreign countries (Meyer, 2003). Although a foreign direct-investor can be an individual, a group of individuals, or a government, most FDI is made by public and/or private enterprises, incorporated or unincorporated. FDI can take place either through the direct entry of foreign firms or the acquisition of existing domestic firms made with the objective of obtaining lasting interest. The economics literature distinguishes between different forms of investment or international money movement, and foreign investment falls into the two main categories of FDI and portfolio investment (Hlaing and Kakinaka, 2019). Nonetheless, Fan (2002) argues that neo-classical economists have

DOI: 10.4324/9781003155492-2

been unable to differentiate FDI from portfolio investment, and contends that the impact each has on a country's development has led to inadequacy in explaining FDI.

There has been years of arguments in favour of and against FDI, beginning with Dunning and Kundu (1982). It has been argued that FDI can lead to economic growth and productivity increases as a whole (Andergassen and Candela, 2009), although the risks associated with FDI have also been emphasised (Lee and Chen, 2020). Tourism FDI has emerged as an important factor in achieving sustainable economic growth for both developed and developing countries, the reason being that the tourism sector requires goods and services, in addition to knowledge, infrastructure, transportation, financial and capital accumulation, and access to international distribution channels. The role of the state in helping to secure positive local impacts from Trans National Corporations (TNCs) and FDI is noted by Liu and Dicken (2006, p.1229) who argue that to gain:

> *obligated embeddedness on the part of TNCs—and for the state and its citizens to gain its benefits—the state not only has to have the theoretical capacity to control access to assets located within its territory, but also the power actually to determine such access.*

Governments in tourism economies often place high priority on attracting FDI due to the tourism sector being a potential path for economic and human development. This is not surprising considering the numerous economic and social benefits FDI offers to both host destinations and tourists. The growth of tourism arrivals stimulates an increasing demand for goods and services. In order to further expand domestic products and infrastructure, to accommodate this growth, governments often prefer to attract FDI to cover the tourist demand for goods and services

The question of who benefits from FDI has stimulated research interest but the challenges of measuring FDI have been noted by researchers, including Fujita (2008, p. 1):

> *The quality of the data available for analyzing foreign direct investment (FDI), particularly in developing countries, does not often meet the required standard for the purpose of rigorous policy analysis.*

Despite difficulties relating to the availability of data and its quality, an interest in local benefits is apparent in the work of many FDI researchers. In this respect the integration of or linking of TNCs with national economic systems has been recognised to be highly desired. As Pavlinek (2004, p. 54) remarks:

> *Linkages of foreign-owned plants with domestic firms are considered the most important mechanism through which technology transfer takes place, additional jobs are generated, and new local enterprises are formed.*

One particular concern that has been raised is that a lack of local linkages "make it relatively easy to move production to lower cost locations should the need arise" (Pavlinek, 2004, p. 52). The notion of destination substitutability in tourism and a lack of loyalty from TNCs (as well as tourist markets) is one that is well-acknowledged. However, transnational tourism is often seen as the answer to economic, social, and environmental development due to its significant economic activity with 1.5 billion international tourist arrivals in 2019 (UNWTO, 2020). Tourism development, if properly developed and supported, can be a "quick-win" in overcoming prevailing economic and social conditions. Tourism-related FDI (TFDI) has become a key component of national development strategies for many countries and the growth of TFDI has been phenomenal.

Book Aim and Content

The aim of this book is to bring together a set of cutting-edge research and empirical case studies from international sources. This book considers TFDI as a component of destination management at all levels of a destination and involving a range of stakeholders. It aims to demonstrate current knowledge on TFDI and to provide insights into conceptual issues. It also seeks to provide specialist perspectives on the state of the art of important aspects and issues within TFDI by studying them in specific case study locations.

The book is aimed at postgraduate students, researchers, and academics. It is also aimed at community economic development and tourism officials in communities that are contemplating the development of tourism with foreign investments. Boards of directors or governors of tourism and specifically hotels will also benefit from this book. As will funding and other governmental or NGO agencies involved with the support of tourism, as well as directors, managers, and employees in hospitality and tourism.

The four foundation chapters in Part II provide the context for the book as well as consider critical concepts and influential factors in TFDI. Tourism FDI issues in specific spatial locations are illustrated in Part III, while Part IV concludes thoughts and gives an overview of current issues and future research directions.

Scope and Organisation of This Book

This book is the first volume examining tourism FDI. It focuses on TFDI and fills a gap in the body of literature on FDI and more specifically, on the issues, challenges, and prospects of tourism FDI.

Through a collection of chapters focusing on principles, practices, and diverse country case studies, the book considers TFDI from the economic, social, environmental, and regulatory perspectives. Thus, the book advances understanding of the positive and negative impacts of TFDI as well as how emerging problems and wider implications for local economies and communities can be managed more sustainably. The book also discusses the transformative opportunities offered by TFDI regarding key economic, social, environmental, and regulatory issues.

Chapter 2 opens Part II by exploring the potential advantages and disadvantages of TFDI and investigates the role of policies involving FDI involvement in tourism, which are examined within the framework of overall destination development strategies. By weighing up the socio-economic benefits that result from TFDI against the possible associated costs and impediments to destination sustainable development, this chapter lays an important foundation for the appreciation of TFDI research at different levels of a destination in this book. Chapter 3 has a similar role in this book. It argues that the need to consider the effects of TFDI on host populations and that local resident perceptions must be acknowledged to provide a more holistic picture of TFDI impacts. Chapter 3 is one of the relatively few studies of local perceptions of TFDI. The authors use theoretical approaches to examine local resident perceptions of tourism per se which are discussed and applied specifically to TFDI impacts. It concludes by emphasising the value of intersecting economic theories and the perceptions of affected local communities in TFDI analyses. Chapter 4 is a dynamic investigation of TFDI and sustainable development. It applies a vector error correction model, catering for dynamic, endogeneity, and causality issues to investigate the link between tourism and FDI in the small island economy of Mauritius. Chapter 5 is the final chapter of Part II on the theme of the foundation and sustainability of TFDI. It critically examines TFDI in the context of Kyrgyzstan, Central Asia through an evolutionary lens. It uses the development path of a particular TFDI to explore how it might inform a wider understanding of sustainable tourism development. This chapter is also concerned with issues of governance, political stability, and institutional capacities and highlights these as being greatly influential for the TFDI's development trajectory alongside collaboration and negotiation processes.

Current issues and challenges in TFDI are the theme of Part III, which comprises eight case study chapters. Chapter 6 uses a panel gravity model to investigate the bilateral FDI stock inflows between France and 19 investor countries in the Hotel and Restaurant industry from 2000 to 2019. As such the chapter shows that bilateral FDI stock inflows between France and investor countries are positively affected by their income and are inversely proportional to the distance between them. Chapter 7 examines why Croatia still needs FDI in tourism and examines the effects of previous FDI in tourism by mapping the effects from the micro to the macro level. It unequivocally points to the structure of accommodation capacities in relation to other EU member states and the low representation of hotel accommodation. Chapter 8 examines the link between FDI, tourism growth, and economic growth of 41 Sub-Saharan African (SSA) countries by using panel data set from 1995 to 2019. It investigates the short- and long-run effects of

FDI and economic growth on the development of the tourism industry using the autoregressive distributed lag (ARDL) model. Chapter 9 employs various econometric techniques to examine the relationship between FDI and tourism, measured by international tourist arrivals and international tourism receipts, in China. The results provide some useful insights for tourism and FDI promotion policies. Chapter 10 is an in-depth exploration of natural sites in Bangladesh and proposes a model that explains and predicts how economic development can be viable through the growth of nature tourism. Chapter 11 is one of the few empirical studies on the nexus between TFDI and economic growth in the case of Indonesia. In order to examine the benefits of TFDI and its effect on economic growth it applies three absorptive capacities, which are human capital, trade openness, and financial market development. The research presented in Chapter 12 is situated in the Caribbean island of Barbados. This is one of the relatively few studies of TFDI in the Caribbean and Barbados and takes a unique approach by critically considering TFDI impacts by focusing on the experienced "realities" as perceived by local hotel employees. It critically reflects on the roles of perceived realities in tourism and FDI. Finally, Chapter 13 provides a summary and conclusion as well as identifies further research needs.

Conclusion

This book aspires to address the significant gap in TFDI research. It is a varied collection of chapters with progressive research relating to TFDI. The combination of theory and case studies is intended to demonstrate our current knowledge as well as identifying research gaps still in existence. As such, this book goes beyond destination-specific content and delivers analytical insights, theoretical advances, and concluding thoughts on further related research areas and needs. The authors and editor anticipate that academics, students, and practitioners will find it inspiring as well as useful, and that the content will stimulate further discussion and research around this important under-researched area.

References

Andergassen, R. and G. Candela, 2009. Less developed countries, tourism investments and local economic development. *Working Paper No. 676*, Bologna: Department of Economics, University of Bologna.

Dunning, J. and Kundu, S.K., 1982. *Impact of Transnational Hotel Chains and Management Companies Operating in Developing Countries u Transnational Corporations in International Tourism, United Nations.* New York and Geneva: United Nations.

Economy Watch (2010). *Definition of Foreign Direct Investment.* Retrieved from https://www.economywatch.com/definition-of-foreign-direct-investment-fdi.

Fan, E.X. (2002). Technological spillovers from foreign direct investment - A survey, *ERD Working Paper Series No. 33*, Asian Development Bank.

Fujita, M. (2008). A critical assessment of FDI data and policy implications. *Transnational Corporations, 17*(2), pp. 107–126.

Hlaing, S.W. and Kakinaka, M., 2019. Global uncertainty and capital flows: Any difference between foreign direct investment and portfolio investment? *Applied Economics Letters, 26*(3), pp. 202–209.

Lee, C.C. and Chen, M.P., 2020. Do country risks matter for tourism development? International evidence. *Journal of Travel Research*, p. 0047287520954539.

Liu, W. & Dicken, P., (2006). Transnational corporations and 'obligated embeddedness': Foreign direct investment in China's automobile industry. *Environment and Planning A, 38*(7), pp. 1229–1247.

Meyer, D. (2003). Outbound UK tour operator industry and implications for pro-poor tourism in developing countries. *PPT Working Paper No. 17*, September.

OECD (1996). *OECD Benchmark Definition of Foreign Direct Investment*. 3rd Edition. France: OECD Publication.

Pavlinek, P., 2004. Regional development implications of foreign direct investment in Central Europe. *European Urban and Regional Studies, 11*(1), pp. 47–70.

UNCTAD (2006). *Developing Sustainable Tourism through FDI promotion*. New York and Geneva: United Nations.

United Nations World Tourism Organisation (UNWTO), 2020. International Tourism Growth Continues to Outpace the Global Economy. Available at: https://www.unwto.org/international-tourism-growth-continues-to-outpace-the-economy

Part II

Current Issues in Achieving Sustainable Foreign Direct Investment in Tourism

2 Foreign Direct Investment in Tourism

Host Destination Opportunities and Challenges

Larry Dwyer

Introduction

Policy makers in many countries view Foreign Direct Investment (FDI) as a means to boost their country's overall pace of development. FDI is considered to create employment opportunities and generate increased national income by enabling build-up of physical capital, developing productive capacity, enhancing skills of local labour through technology transfer and managerial know-how, and promoting economic diversification, while better integrating the domestic economy with the global economy. For countries with developing tourism sectors in particular, FDI is seen as having a potentially very significant role in improving infrastructure such as international airports, highways, hotels, and modern technologies that comprise the basic foundations of tourism development. FDI in tourism, however, presents special challenges and concerns regarding its effects. As we shall argue below, FDI is associated with negative as well as positive effects on the tourism industry of a host destination. Consequently, policies affecting FDI in a country's tourism industry should be carefully assessed, comparing the benefits of such inward investment to the possible wider socio-economic costs to tourism and other industry sectors over the longer term. This comparison is not straightforward and, as we shall see, there are many gaps in our knowledge of the effects of FDI in tourism in the different destinations that comprise the global industry. The proposed framework for identifying the benefits and costs associated with FDI in tourism should facilitate further research in this area, particularly in destination case studies.

FDI occurs when an investor resident in one country (the source country) acquires ownership in and a significant influence over the management of an enterprise or productive asset in another country (the host). Firms invest to derive the maximum expected return on outlays. They exploit their competitive advantages, often in the form of brand equity, technology, or other proprietary know-how, by utilising them in another country. This may involve creating a new enterprise (greenfield investment) or changing the ownership of an existing enterprise (via a merger and or acquisition). FDI is described as a whole package of resources: physical capital, modern technology and

DOI: 10.4324/9781003155492-4

production techniques, managerial and marketing knowledge, entrepreneurial abilities, and business practices (UNCTAD, 2008; Denisia, 2010).

What is unique about FDI in tourism is the separation of ownership and control that frequently occurs (McGaughey, Raimondos and la Cour, 2020). Tourism comprises a large number of diverse and interlinking activities, making the compilation of standardised FDI statistics in tourism at the international level extremely difficult (Endo, 2006). Further, many countries do not distinguish between domestic and foreign investment in their official statistics. The frequent use of managerial contracts or franchising operations (whereby multinational enterprises or MNEs can assume control over an operation without having committed equity capital) creates a growing gap between the official FDI data and the activities of firms in practice.

UNCTAD (2008) reports that FDI in tourism is still rather low – in both developed and developing countries – compared to the levels of FDI in other economic activities, including other services industries. FDI in tourism is mainly concentrated in hotels and accommodation (Nunkoo and Seetanah, 2018). However, its impact can be substantial, especially when the scope of FDI is widened to include non-equity forms of MNE involvement, such as management contracts that are much more prevalent.

Destination Ability to Attract Tourism FDI

The ability of a tourism destination to attract FDI is influenced by a number of factors (UNCTAD, 2008; European Commission, 2016; Nunkoo and Seetanah, 2018).These include the economic growth rate of the host destination; the productivity and performance of the tourism industry, including the profitability of tourism related enterprises; access to sources of financing and the availability of sufficient information and data needed to make investment decisions; government regulations, procedures, and incentives regarding local and foreign investment; the state of the destination physical and social infrastructure; market characteristics; visitor tastes and preferences, expected growth in visitation, transport links, etc.; the ability of the destination to market and promote itself effectively; political and macroeconomic stability of host destination (Dwyer and Forsyth, 1994; European Commission, 2016; Nunkoo and Seetanah, 2018). Only if these factors combine to allow investors to earn an adequate return on their investment will they make the necessary long-term commitment to allocate funds to a host destination (Groh and Wich, 2012).

Motives for FDI in Tourism

The various motives for foreign investment in tourism can be explained in large part using the widely accepted "eclectic paradigm" of international production (Dunning, 2002; Pathan, 2017; Gudowski and Piasecki, 2020). The eclectic paradigm asserts that the extent, pattern, and growth of value-adding activities undertaken by MNEs outside their home countries are dependent on the value of and interaction between three main variables: ownership advantages; location advantages; and market internalisation advantages.

Ownership Advantages

These include the competitive advantages that foreign firms possess in supplying any particular market or set of markets. These advantages may arise from firms' privileged ownership of, or access to, specific technological, managerial, financial, or marketing assets. Ownership advantages may be of a structural or behavioural nature. For the tourism industry they might include: the size of the company and its ability to obtain economies of scale; proprietary competitive advantages such as strong brand name or established reputation in providing tourism services that allow the firm to increase market penetration; greater availability of equity finance to support tourism industry expansion; better knowledge of, and favoured access to, international tourism markets; better-trained personnel; better management, and reservation systems; and better organisational and IT capabilities to successfully integrate separate value-adding activities (Snyman and Saayman, 2009; Park and Xiao, 2017).

Location-Specific Advantages

When capital can move freely from country to country, it is relatively straightforward for firms to locate and invest abroad, and repatriate profits. The advantages of location are the benefits of value-adding activities that combine ownership-specific advantages with immediate factor endowments in a foreign country (Snyman and Saayman, 2009). Factors that make a destination attractive to international investors include: size, growth, and stage of development of the overall tourism market in the host country; existing tourism facilities in the host country, including type of attractions, appealing environmental and heritage features; destination market potential in the context of global and regional tourism trends; openness of the host government towards FDI in the country including investment incentives; political, social, and economic stability of the host destination; quality of general infrastructure in destination (transport, telecommunications, water, power); input prices, quality, and productivity; lack of entrepreneurship in the host country, reflected in an unwillingness to take risks to develop tourism facilities; deficiencies in the host country's capital market as it affects tourism, with associated barriers to local investors; lower taxation rates in the host destination compared to alternative destinations; compatibility with the firm's strategy to diversify operations for the purpose of reducing risks geographically (Assaf, Josiassen and Agbola, 2015; Li, Quan, Stoian and Azar, 2018).

Market Internalisation Advantages

Market internalisation advantages refer to the advantages of a firm controlling and coordinating ownership and location-specific advantages within an MNE hierarchy, rather than selling the right to use those advantages to domestic firms in the host country. In tourism, these advantages might include: easier control by a firm over the character of the tourism product, including security of supply, price, and quality; better planning opportunities; better coordination opportunities; greater opportunities to increase market share and profits (Gaur, Pattnaik, Singh and Lee, 2019).

The interrelations of ownership, location, and internalisation advantages, and the response to them by firms, vary according to industry, countries of origin and destination of investment, and firm specific characteristics (Groh and Wich, 2012). They also vary over time as changes in technology and the entrepreneurial and economic environment affect the competitive position of firms and the location of their value-adding activities. The utilisation of these advantages depends primarily on the relative costs of equity and non-equity forms of managing interrelated economic activities (Dwyer and Forsyth, 1994).

Effects of FDI on Host Destination

FDI potentially has both positive and negative effects on the host destination. In its positive effects, FDI may foster productivity gains, technology transfers, introduction of new processes, managerial skills and know-how in the domestic market, employee training, international production networks, and access to export markets (Denisia, 2010; Pathan, 2017). FDI can be important to any economy, especially to developing nations increasing the opportunities for global networking. Negative effects may include crowding out of domestic investment, greater "leakages" from revenue due to production inputs sourced externally, including repatriation of profits, limited opportunities for career advancement of local employees, inappropriate form and scale of industrial development, loss of equity in and control of the industry development, economic dependence on foreign investors, and adverse environmental effects. Each potential type of impact is associated with FDI in tourism.

Potential Advantages of FDI to Host Destination

FDI in tourism has the potential *inter alia* to increase tourism exports, contribute to growth and development; create jobs; build exports; provide additional sources of finance for tourism expansion; increase opportunities for global marketing and networking, improve tourism product quality; and facilitate technology transfer and innovation.

Increased Economic Growth in Host Destination

Given shortages of facilities and infrastructure in many destinations, FDI is considered an effective channel for transferring the trade, knowledge, and technologies leading to economic growth. Tourism-related MNEs often establish linkages with local suppliers and distributors that can boost the host country's economic activity and business opportunities (Nunkoo and Seetanah, 2018). When FDI is directed towards the least developed countries and associated with the creation of employment for unskilled workers, it has the potential to contribute importantly to poverty-reduction (Dwyer and Čavlek, 2019).

It has been argued that FDI has a larger impact on growth than other international capital flows – such as portfolio investment and bank loans – because of the limited volatility of FDI. This relates to the fact that FDI cannot easily be withdrawn while profits, losses, and risks are shared among the foreign and the host entity. FDI is thus attracted by the long-term prospects of the country and its policies, and is therefore more stable than other capital investments (Albuquerque, 2003). Other types of external capital are known to be less stable and of shorter term, thereby hindering sustainable growth (Colen, Maertens and Swinnen, 2009).

While FDI can, under certain conditions, be an engine of economic growth, its impact is conditional on factors such as the type of FDI, the economic sector, the size and quality of capital stocks, and the absorptive capacity of the host economy (Andergassen and Candela, 2013; Chingarande and Saayman, 2018). FDI is considered to affect economic growth in a different way than domestic investment because FDI entails – besides the accumulation of physical capital – a bundle of potentially growth-enhancing attributes including technology, managerial knowhow, entrepreneurial ability, and access to global distribution networks and international markets (Dunning, 2002). Through such means, FDI contributes importantly to economic growth and to the overall development of the host country – *directly* through the accumulation of capital, technological know-how, and innovative capabilities (Liu, Shu, & Sinclair, 2009), and *indirectly* through technology and knowledge spillovers to domestic firms in the host economy. However, empirical evidence in support of these links theories differs according to destination and industry context (Colen et al., 2009).

There is ongoing debate as to whether tourism primarily is a catalyst for growth, increasing the GDP of destinations or whether tourism growth is primarily an effect of economic growth in the domestic or international economy. The evidence seems to be that it is both (Brida, Cortes-Jimenez & Pulina, 2016; Yazdi, Nateghian and Rezaie, 2017; Sokhanvar, 2019). The impacts of FDI on a host destination can argued to support the notion of Tourism-led Growth (TLG). The export-led growth hypothesis emphasises that increased exports (including tourism exports associated with FDI) can generate foreign exchange, create employment, stimulate additional investments, help to exploit economies of scale, promote efficient allocation of resource efficiencies, and diffuse technical knowledge (Chingarande and Saayman, 2018).

Not all types of FDI are likely to create substantial additional employment in host economies. When FDI represents additional investment (greenfield investment), it generally provides employment, while mergers and acquisitions are less likely to create additional jobs (UNCTAD, 2008). Moreover, when foreign firms compete with local firms resulting in crowding-out, employment in domestic companies and in the sector as a whole may be reduced (Dwyer and Cavlek, 2019). Hence, the magnitude and sign of the employment effect will depend on the industry of investment, mode of entry of FDI, and destination characteristics (Colen et al., 2009).

Provision of Finance for Growth and Development

The availability of financial capital is critically important for achieving tourism development and economic growth. FDI is assumed to augment domestic capital thereby stimulating the productivity of domestic investments (Denisia, 2010). Globalisation of product and financial markets coupled with deregulation, is associated with increased economic integration in specialisation and economies of scale, resulting in greater trade in financial services through both capital flows and cross-border entry activity and greater financial support for FDI. The development of complex financial products, such as derivatives, has enabled global credit markets to grow rapidly. Risk reduction via diversification can be accomplished through company involvement with international financial institutions and partnering with both local and multinational businesses.

Many destinations and operators have limited or no access to credit to support investments and innovation. Strong FDI in a destination can reduce the risks for domestic investors and provide a catalyst for finance (UNCTAD, 2008). FDI can stimulate economic activity in a destination, enabling residents to enjoy higher economic growth, employment, and standard of living than could be achieved from domestic capital investment alone. FDI provides access to capital to supplement a destination's domestic savings. It helps to unlock capital constraints particularly in lesser developed economies where financial markets are narrow and poorly developed. Effectively, the host country "imports" a service (long-term equity holding) from a more readily available source in another country (Dwyer and Forsyth, 1994; Al-Hallaq, Athamneh and Suleiman, 2020).

Tourism development often requires long-term capital investments, funding for which may not be in adequate supply in a destination, particularly in a developing country. FDI is considered by many researchers to be essential in creating and upgrading tourism-related infrastructure (Chingarande and Saayman, 2018; Nunkoo and Seetanah, 2018). The tourism industry requires capital, infrastructure, knowledge, and access to global marketing and distribution chains (Samimi, Sadeghi, and Sadeghi, 2013). The availability of financial sources is therefore essential for furthering tourism development and economic growth, especially in the case of capital-intensive tourism projects that are often tied up with huge set-up costs. These include international airports, highways, shipping terminals, hotels, tour operators, travel agencies, car rental, restaurants, and tourist attractions, which are critical to the success of tourism (Samimi et al., 2013). However, in countries where governments, in order to attract FDI, extend tax exemptions to MNEs – as is the case in many developing countries – the potential for poverty-reducing effects through tax revenues and redistributive measures are limited (Colen et al., 2009).

FDI Expands the Size of the Host Tourism Industry

FDI and MNE operations have the potential to significantly contribute to the growth of tourism in developing countries (Perić and Radić, 2016). A dynamic, growing economy supported by FDI will increase revenue generation and create employment, leading to higher continued personal and household incomes that can be used to fund tourism experiences, especially domestic and outbound. The increased inbound tourism expenditure can generate additional activity in related industries, acting as a catalyst for their growth (Ghebrihiwet, 2017; Maryam and Mittal, 2020). Increased inbound tourism is argued to lead to business development across the tourism sector itself and along industrial value chains enhancing efficiency through increased competition among firms, facilitating the exploitation of economies of scale in local firms (Chingarande and Saayman, 2018). The increased inbound tourism also generates foreign exchange that can be used to import capital goods in order to produce goods and services, leading to further economic growth (Samimi et al., 2013). Studies show that tourism growth in a destination can affect the level of FDI (Tiwari, 2011; Fang, Gozgor, Paramati and Wu, 2020).

It must be noted however that the size of the multiplier effects from increased inbound tourism expenditure will depend upon the factor constraints. Tourism growth will tend to draw resources away from other industries, inducing a contraction in their production with associated job losses (Kang and Lee, 2011). Consequently, additional inbound tourism associated with FDI may not have the expected economic impacts once industry interactive effects are accounted for (Dwyer, Forsyth and Spurr, 2003).

Several studies have found a positive link between increased FDI in a host destination and economic growth via increased inbound tourism flows including business tourists (Selvanathan, Selvanathan and Viswanathan, 2012; Al-Hallaq, Athamneh and Suleiman, 2020). In particular, FDI is argued to provide additional capacity for small island states, allowing them to expand their tourism activities (Barrowclough, 2007; Jayaraman, Chen and Bhatt, 2014). Overall, findings suggest that FDI positively impacts on destinations and that the relationship between FDI and tourism is of a bidirectional nature (Nunkoo and Seetanah, 2018).

Marketing and Destination Promotion Effects

MNEs play an important role in the economies of both developed and developing countries in that they allow the host countries to be integrated into international tourism networks that lead to an increase in the flow of tourists and generate more income from tourism-related activities (Fauzel, Seetanah and Sannassee, 2017). Market connections related to global marketing and access to global distribution networks can generate greater flows of tourists to those countries that host FDI or MNEs involved in the tourist industry. MNEs link local suppliers and sub-contractors to international markets, provide information on foreign markets conditions and consumer preferences, and offer distribution networks, transport infrastructure, and export management skills (Pathan, 2017). Emerging destinations, characterised by under-resourced national tourism offices, often lack resources to mount sustained marketing campaigns in major potential long-haul markets with low awareness of the destination. Compared to resident investors, the foreign investor may have better knowledge of their home country's travel market and be better placed to market the destination more effectively in that market (Colen et al., 2009). Foreign investment may result in greater or superior promotional effort in the home country of the investor, leading to higher visitor numbers from that country to the host destination. Overall, the involvement of MNEs in marketing and promotion activity can increase destination awareness and inbound numbers from many origin countries. Tourism related MNEs can also contribute towards building and/or reinforcing the positive image of a destination in which they choose to locate. The role of well-known hotel brands with a global marketing reach may be particularly important for developing countries that have limited resources or abilities to promote their destinations, or that have not yet developed their own "brand identity" (Anwar and Sun, 2016).

Product and Quality Effects

In a globalised world, FDI is largely responsible for the wide range of consumer choice in most products that are available in markets today. It allows MNEs to take advantage of global economies of scale in order to expand their operations into new markets to offer consumers products at competitive prices, while also providing opportunities for domestic firms to export to wider markets (Dwyer and Čavlek, 2019). Multinationals may be better able to monitor and control the quality of the services they provide to international tourists. In contrast, small, domestically owned and operated businesses, particularly in developing countries, may have limited knowledge of tourist expectations regarding service standards and may find it difficult to compete internationally. Tourism often has difficulty in attracting or maintaining enough skilled personnel to work in its various sectors due to skills gaps, seasonality, and lack of career opportunities (Dwyer, 2015). This can lead to problems in service quality and erode destination competitiveness. Established tour operators give quality guarantees to the tourist, thereby reducing the perceived risks of default, unsatisfactory service, or other transactional uncertainties. A foreign brand name may act to reduce the information search costs of potential tourists especially in choosing to holiday in developing countries. Foreign brand names tended to enhance destination image among business tourists in particular (Anwar and Sun, 2016).

Technology Transfer

FDI can increase the existing stock of knowledge of the host country through labour training, transfer of skills, and the transfer of new managerial and organisational practice. Technology encompasses not only the "hardware" of building and design of hotels and restaurants, but also the "software" relating to skills developed. One of the main reasons for (and benefits from) foreign investment is the transfer of soft technology including managerial expertise and links to international value chains. FDI can contribute to the formation of human capital – resulting in spillover effects to the rest of the economy – both by demanding and by supplying skills critical for development. The availability of workers with appropriate skills is a key requirement for the successful development of a tourism industry. FDI encourages local businesses to innovate and keep up with global best practices (Pathan, 2017). FDI brings technological improvements associated with establishing new businesses or modernising established ones in a destination. MNEs can increase the productivity of labour by supplying foreign technology and improved training methods. MNEs can thus make a significant contribution to the development of human capital which remains critical for tourism development (Dwyer, 2015).

Foreign firms assist in technology and knowledge transfer, involving also a "demonstration effect" on local entrepreneurs. Technology, especially ICT, is fundamental to prosperity in the knowledge economy, where the skills to transform knowledge and information into innovative products and services are a defining success factor (UNCTAD, 2008). In developing destinations in particular, increased skills levels of management have been essential in catering to the demands of foreign tourists and in maintaining the international competitiveness of the local product. FDI in tourism, for instance, can also contribute significantly through the adoption of a diverse range of new technologies and skills that may lead to substantial technological and skills diffusion into the economy (García, Jin and Salomon, 2013). An increased demand for skills is expected to raise the wage and employment opportunities of skilled workers, creating incentives for overall investment in human capital (Iwasaki and Tokunaga, 2016).

The transfer of knowledge and skills is considered as the most important contribution of FDI in the hotel sector by developing and industrialised countries (UNCTAD, 2008). International hotel chains have substantial in-house training programs emphasising systems of accounting, procedures, and management. Their operations have resulted in a substantial increase in skills levels within the hospitality industry worldwide. Such training programs are a form of "technology transfer", creating spillover benefits to domestic firms desiring access to a more skilled pool of labour within the tourism sector.

While the above arguments support the role of FDIs in technology transfer, it should be emphasised that developing destinations may fail to meet enabling conditions for strong technology transfer. A barrier to the ability of developing countries to gain the full benefits of technology transfer is that they may not have the requisite infrastructure in place to leverage technological benefits. The absorption of IT and its benefits is not automatic. Many destinations will fail to meet the conditions needed for effective IT utilisation – high educational levels, adequate infrastructure, and appropriate regulatory policies. Destinations with more developed human capital and financial systems have better chances of achieving growth through technology transfer (Feng, 2020).

It is often argued that FDI contributes directly – and more strongly than domestic investment – to higher levels of growth in an economy due to the more advanced levels of technology, managerial capacity, and know-how, resulting in higher levels of efficiency and productivity (Colen et al., 2009). Indeed, the superior position in terms of technology and know-how of MNEs has been underlying many of the arguments in favour of FDI liberalisation policies (Ghebrihiwet, 2017). However, the assumption of foreign firms being more efficient than domestic firms is not necessarily true. The inflow of capital might not always be accompanied with improved technologies, managerial capacity, and entrepreneurial ability. Foreign investment can take place because foreigners have a superior cash position and can take advantage of liquidity–constrained domestic investors' fire sales, rather than because of a technological advantage (Krugman, 2000).

Potential Disadvantages of FDI to Host Destination

Many activities associated with FDI may actually impoverish weaker groups through displacement of locally owned businesses, increased economic dependence on decisions of external investors, loss of access to resources, cultural disruption, adverse social and environmental impacts, and so on. Some important potential negative effects include the following.

Crowding Out of Domestic Investment

As a consequence of FDI, domestic owners of facilities may lose through lower prices and hence lower expected rates of return on their investments.

In more capital-intensive tourism-related sectors (such as large hotels, transportation) it may be far more difficult for domestically owned firms to compete with large inward investors. If local investors are crowded out by foreign investors, the effect of allowing foreign investment may be to reduce tourism net investment. This is particularly likely to occur where the local investment pool is small. As highlighted above, tourism expansion may lead to de-industrialisation (Kang and Lee, 2011). To the extent that this occurs, domestic investment may fall to lower levels in some non-tourism contracting sectors also (Dwyer and Forsyth, 1994).

In sectors where the existing range of facilities is limited and of low quality, crowding-out effects may be substantial. In contrast, where certain investment opportunities (for example, luxury resorts), have not been explored sufficiently by domestic sources, the potential for crowding out of domestic operations is likely to be less. For countries with substantial domestic use of their tourism industries (such as France or USA), the losses faced by resident investors are likely to be made up by gains enjoyed by domestic tourists. However, for small countries that rely heavily on foreign tourism and have limited supplies of local equity, foreign investment could impose a net cost, especially if a high proportion of the resources used in tourism (for example, land near beaches) are foreign owned. This needs to be compared to the gains achieved from increased tourism expenditure (Dwyer, 2015).

Many destinations provide a range of financial, fiscal, and other incentives to induce foreign investment in the tourism sector. When both foreign and local investors are eligible for government subsidised loans, local investors may be crowded out. The extent of crowding out depends on the type of destination (Dwyer, Forsyth, Madden and Spurr, 2000; Kudina and Pitelis, 2014). However, since financial incentives must inevitably be paid for by local taxpayers, increased incentives offered to foreign investors can result in lower gains to host countries. Policies offering preferential treatment and incentives to attract FDI – such as export free zones and other tax incentives – may introduce a distortion that negatively affects domestic investment and limits growth spillover effects through crowding-in (Sharma, 2016).

A related risk associated with FDI involves over-specialisation, such as being over-reliant on producing a limited range of goods for particular tourism markets. Over-reliance on tourism, especially mass tourism, carries significant risks to local communities. Two major types of over-specialisation may be identified. One type relates to a dependency on tourism in general as an export market. Emerging tourism destinations in particular, are vulnerable to any disruptions that occur in the developed countries that purchase their products, as well as natural disasters that deter tourism inflows. The global financial crisis and the COVID-19 pandemic both demonstrate the risks involved in tourism dependency arising from sudden unfavourable changes in demand from world markets (Sheldon and Dwyer, 2010). Another type of over-specialisation occurs within the tourism market. Over-reliance on tourism from particular origin markets or on a particular tourism product (e.g., conventions, health tourism, gaming). Changes adversely affecting these markets, making them less competitive and adaptable, can result in substantial costs to tourism destinations that develop a dependency on them. As a counter to this, MNEs may have "deeper pockets" than locally owned firms, and thus may be better able to operate during some tourism downturn maintaining the provision of facilities to visitors. There is some evidence that MNEs bring market stability and confidence to a host tourism industry experiencing a crisis (Barrowclough, 2007). Given that many lesser developed economies are vulnerable to fluctuations in world commodity prices, more attention must be given as to how destinations can participate in new forms of tourism to avoid the types of dependency that characterise existing forms.

Greater "Leakages" from Tourism Imports

FDI and the rise of MNEs within increasingly liberalised trade systems have sped up processes of off-shoring making it difficult for local destinations to keep the money spent by tourists within their localities. The extent of leakages from tourist expenditure and business profits depends on how strongly the tourism sector is integrated into the destination economy through backward and forward linkages with other sectors and integration into regional and global value chains (Dwyer, 2015; Dwyer and Čavlek, 2019). FDI itself may be directly associated with higher use of imported goods, materials, and foreign expertise according to the standards and tastes of developed countries, reducing the (initial) net foreign exchange earnings that might accrue in the case of local ownership. However, two qualifications apply: First, The size of the additional leakages depends importantly on how foreign owned facilities source their inputs (land, labour, capital) as compared to domestically owned facilities. The source of inputs (whether domestically produced or imported), and the size of the resulting leakages, will help to determine the income generating effect of the investment itself and tourism expenditure associated with its operation. Second, even if additional imports are associated with the FDI, this puts downward pressure on the exchange rate, encouraging exports and discouraging imports in other sectors (Dwyer and Forsyth, 1994).

Determination of any additional outflows due to foreign investment in tourism facilities is not straightforward. In determining incremental foreign exchange outflows resulting from foreign investment in the tourist industry, one needs to be specific as to what would have happened if this investment did not take place. The extent of any differences in leakages depends on whether the foreign investment replaces *similar* domestic investment, replaces *different* domestic investment, or *adds* to overall investment in the tourist industry. The effect on economic activity resulting from FDI will be different in each case. FDI does not, in itself, imply greater use of foreign sourced inputs. The extent to which foreign, rather than domestically sourced inputs are employed depends on their relative costs, quality differences, and local availability (Dwyer and Forsyth, 1994). While the high import content of goods used to satisfy tourist needs generally is recognised for lesser developed economies, it is the destination's industrial structure, rather than ownership of tourism facilities, that determines the degree of import penetration. An exception to this is the incidence of tourism resort "enclaves" which may result in minimal economic benefit for the destination due to dependence on international charter operations, expatriate employees, and imported food and other equipment. In such cases, there is a high leakage rate (Brohman, 1996).

Another form of leakage arises when management fees and profits associated with FDI flow out of the host country reducing the economic contribution from tourism expenditure. Many critics go further and view repatriated profits as a "leakage" from the host economy due to the foreign investment. This view rests on a misconception. While a proportion of the sales of a foreign owned tourism facility will go towards profits that can be repatriated to the owners, these profits are not "lost" to the host destination in any meaningful sense, since there would be no profits to repatriate had the investor not first brought the capital to invest (Dwyer and Forsyth, 1994). If profits paid overseas are regarded as a "leakage" from the economy, then the initial payment for the facility should be regarded as an "injection" of money that would not have occurred except for the foreign investment.

The leakages issue must be analysed in the context of the broader potential for tourism development to generate greater income and employment in the host destination (Dwyer and Forsyth, 1994). Even if foreign ownership of a facility leads to a more direct input-sourcing from overseas or repatriation of profits, the economy might gain more in income and jobs than it otherwise would from other forms of investment. The expanded size of the tourism industry may also lead to more reliance on domestic sources of inputs by competing facilities. Where domestic firms are crowded out or play a small role in the tourism supply chain, policy measures can be enacted to forge greater links between foreign owned facilities and local suppliers so that the latter can capture more of the tourism value chain. Local procurement may also offer wider benefits in terms of a destination's ability to market itself.

Growing Income Inequality

There is substantial debate as to whether economic growth improves or reduces income distribution among destination residents (Alam and Paramati, 2016; Fang et al., 2020). A causal relationship between FDI, inbound tourism, and economic growth (see above) may not necessarily lead to higher living standards in many destinations, particularly in developing countries. This depends on the existing distribution of income and wealth associated with factor ownership and the quality of services such as health care and education (Blake, Arbache, Sinclair and Teles, 2008). The empirical literature also shows mixed results on the impact of FDI on inequality (Colen et al., 2009). A widely accepted view has been that, over time, as nations become wealthier and more efficient, the benefits of trade will "trickle down", reducing poverty levels. This "trickle down" theory has been criticised as unrealistic (Akinci, 2018). FDI can help to reduce income inequality when its benefits favour the poor and those in the lowest income categories. Except for "socially responsible" investment in poverty-reducing projects out of charity or image-building, there is no direct link between FDI and poverty (Nunnenkamp, 2004).

Tourism expansion may have a significant negative impact on income inequality in developing economies (Fang et al., 2020). The benefits of tourism development may only be confined to an elite class of people in society, such as the owners of the tourism service providers, entrepreneurs, investors, and managers of tourism enterprises. In addition, tourism-related business enterprises may create only low salaried jobs in the local communities and exploit their services and resources, which eventually increase the income inequality in any given society (Alam and Paramati, 2016). There is also evidence that FDI may cause unemployment in industrialised countries as businesses outsource work to developing countries where the cost of labour is low. This undoubtedly affects the more developed economies that comprise some of the main tourism origin markets. Outsourcing, with its potential to result in structural unemployment, can lead to domestic job-losses and lower quality of service (Dwyer, 2015)

It is sometimes argued that employment opportunities especially in managerial positions for locals may be limited owing to the use of the expatriate labour with limited opportunities for career advancement of local employees (Brohman, 1996). In particular, in order to maintain firm specific advantages, it is claimed that key management positions may be held by expatriates, and that only lower-level personnel requiring low skills are trained for reasons of service quality and performance. The major reason for imported labour is to maintain quality in the provision of tourism services. Where imported labour serves to generate revenues and to enhance the quality of the destination's tourism products, including customer service, there may be wider benefits by way of tourism's contribution to GDP and foreign exchange earnings, not readily apparent from the examination of local employment shares. Where FDI increases destination tourism facilities, expatriate staff can hardly be regarded as taking jobs from locals since many of the jobs very likely would not exist but for the foreign investment. Even in cases where some employees are foreign citizens, they will pay local income tax out of their wages and will meet their living expenses within the destination, injecting expenditure into the economy with positive economic effects. MNEs often provide a systematic training program to staff, directed at international markets, in addition to international placements within a firm's different locations. This has the attendant "technology transfer" benefits to the host economy as discussed above. In any case, governments and destination managers have a wide range of policy options that can be enacted including the provision of programs to develop an indigenous workforce with skills required by the tourism industry.

In many developing countries, FDI may cause local economic activities and resources to be used less for the benefit and development of local residents and communities and increasingly for export and the enjoyment of others (foreign tourists). This occurs in tourism when the domestic market is neglected. To avoid overdependence on the international tourism market and related problems, destinations can consider tapping the potential of domestic tourism. In geographically large destinations in particular, domestic tourism can effectively absorb the excess supply resulting from the slump and seasonality of inbound tourism. In addition, an expanding domestic tourism industry helps redistribute the national income, thus reducing interregional gaps in the level of economic development and contributing to social equality. Typically, the domestic tourism segment has much lower entry barriers and broader and deeper linkages with local economy as well, thus greatly facilitating local participation. Essentially, a vibrant domestic tourism industry provides the foundation for a competitive international tourism sector (Dwyer, 2015). A prudent strategy for destinations as a whole and its local communities is to diversify its economic base as well as the range and appeal of its tourism products. This strategy would involve calls for more local rather than foreign involvement in tourism investment.

Inappropriate Form and Scale of Tourism Development

The concern is sometimes raised that foreign owned tourism facilities are usually of greater scale than domestically owned facilities (Barrowclough, 2007). Mass tourism may be regarded as one of the most visible manifestations of FDI in tourism. The characteristics of mass tourism associate with FDI in tourism tend to favour the development of large scale integrated facilities and that this may be inappropriate for smaller, less developed destinations. Thus, multinational hotel companies may affect the scale (size) and type (class) of hotels of hotels constructed, thus affecting the general scale of tourism development. Large scale foreign owned enclave type facilities do destroy valued environments and alienate local populations (Brohman, 1996). But so also do large scale domestically owned facilities. Nevertheless, the larger size of foreign owned facilities tend to involve greater exploitation of scarce environments and increased demand for resources such as power and water. In terms of social impact, critics point to the rising alienation of local populations, persistent socio-economic inequalities, and loss of cultural identity associated with MNE operations (Nunkoo and Seetanah, 2018).

Some researchers, however, emphasise the importance of the global information network in driving the "social corporate responsibility" idea of multinational companies. Foreign investors bring not only capital and technological know-how, but also a specific corporate culture of setting higher social and environmental standards for its operations, compared to local competitors. Therefore, over time, FDI would be a force for raising standards in developing countries (Colen et al., 2009). Similar arguments have been made regarding the environmental effects associated with FDI. While multinationals have been accused of investing in developing countries to take advantage of low environmental regulations (the pollution haven hypothesis), they are more recently seen as leaders in the introduction of good environmental practices, cleaner energy, and "green technologies" into developing countries (Tamazian, Chousa and Vadlamannati, 2009). In respect of social impacts, FDI in the tourism sector has been found to play a key role in promulgating and expanding indigenous tourism sectors across a wide spectrum of developing nations including the small island economy of Mauritius (Fauzel et al., 2017).

If a destination wants to develop alternative small scale locally owned managed and integrated tourism in some regions because of the small size of the country and fragile environmental or socio-cultural concerns, it should carefully evaluate what role, if any, MNEs can play in that type of development. However, the problem seems not to lie so much in the "foreignness" of the ownership but in the nature of the planning and zoning laws that permit such constructions. Good environmental management by local communities is important for all forms of tourism development.

Loss of Equity and Control

Foreign ownership of tourist facilities implies that a host nation loses some equity in its tourism industry and that the conduct of firms in the tourism industry is more vulnerable to decisions made outside the host destination. A related concern is that developing countries can become over-dependent on MNEs given the market connections enjoyed by international tour operators, hotel companies, and airlines (Brohman, 1996; Nunkoo and Seetanah, 2018). Some would argue that MNEs increasingly are able to influence intergovernmental negotiations and rule-making, with continuing erosion of democracy and community control (Elkomy, Ingham and Read, 2016). The concern is that host destinations may be placed in a dependent, vulnerable situation because the size and type of international tourism demand is determined by the actions of people in wealthier countries. A related concern is that MNEs are increasingly "footloose", with the ability to switch their investments between territories in search of the most favourable regulatory regimes (Krugman, 2000).

While foreign owned firms obviously do exercise their power in the international marketplace, the issue comes down to the long-term contribution that FDI makes to the host destination as compared with its absence. This must be determined on a case by case basis. As a highly interconnected industry, maximising the potential of tourism requires coherent and comprehensive policy frameworks to ensure that sustainable tourism is integrated into the country's overall economic, social, and environmental policies. To firm up their control over MNE activities, destinations that seek to access FDI in tourism as part of their overall economic development need to design and implement appropriate policy frameworks that will improve the bargaining position of their domestically owned tourism service suppliers.

Destination governments and agencies at all levels can negotiate with MNEs regarding the terms of involvement in their economies and thus align them with national policies, including engagement in financing infrastructure and operating services. In this way, FDI, in its catalytic role, can be "managed" cooperatively by the investment sources and the host government so as to achieve sustainable tourism development. Ultimately, host governments can exercise whatever policies are needed to ensure compliance (Dwyer, 2015).

Conclusion

While economic theory predicts a positive impact of FDI on economic growth, empirical evidence is mixed. FDI has the potential to help address some of the challenges facing destinations in an increasingly interdependent, globalised world. Governments reserve considerable power to determine how the industry performs. Policies can be enacted to increase local economic participation in order to reduce economic dependencies on other countries. Given that FDI is associated with both positive and negative effects on the tourism industry of a host destination, one should be cautious in claims regarding its potential as a catalyst for tourism industry and economic development. The extent of costs and benefits of FDI to a host destination will vary considerably from case to case. Any policies involving FDI and MNE involvement in tourism should be examined carefully within the framework of overall development strategies, weighing up the socio-economic benefits that result against the possible associated costs to destination sustainable development.

References

Akinci, M. (2018) 'Inequality and economic growth: Trickle-down effect revisited'. *Development Policy Review, 36*, pp. 1–24.

Alam, M.S. and Paramati, S.R. (2016) 'The impact of tourism on income inequality in developing economies: Does Kuznets curve hypothesis exist?'. *Annals of Tourism Research, 61*, pp. 111–126.

Albuquerque, R. (2003) 'The composition of international capital flows: risk sharing through foreign direct investment' *Journal of International Economics, 61*(2), pp. 353–383.

Al-Hallaq, S.S., Athamneh, A.B., and Suleiman, H.M. (2020) 'The impact of foreign direct investment on the growth of the tourism sector in Jordan (1980–2016)'. *Journal of Public Affairs, 20*(2), e2005.

Andergassen, R. and Candela, G. (2013) 'Less developed countries, tourism investments and local economic development', *Review of Development Economics*, *17*(1), pp. 16–33.

Anwar, S. and Sun, S. (2016) 'Foreign direct investment and product quality in host economies'. *Stanford Centre for International Development Working Paper No. 550*. Available at SSRN 2529883.

Assaf, A.G., Josiassen, A. and Agbola, F.W. (2015) 'Attracting international hotels: Locational factors that matter most'. *Tourism Management*, *47*, pp. 329–340.

Barrowclough, D. (2007) 'Foreign investment in tourism and small island developing states'. *Tourism Economics*, *13*(4), pp. 615–638.

Blake, A., Arbache, J.S., Sinclair, M.T. and Teles, V., 2008. Tourism and poverty relief. *Annals of Tourism Research*, *35*(1), pp. 107–126.

Brida, J.G., Cortes-Jimenez, I., and Pulina, M. (2016) 'Has the tourism-led growth hypothesis been validated? A literature review', *Current Issues in Tourism*, *19*(5), pp. 394–430.

Brohman, J. (1996) 'New directions in tourism for third world development'. *Annals of Tourism Research*, *23*(1), pp. 48–70.

Chingarande, A. and Saayman, A. (2018) 'Critical success factors for tourism-led growth'. *International Journal of Tourism Research*, *20*(6), pp. 800–818.

Colen, L., Maertens, M., and Swinnen, J. (2009) 'Foreign direct investment as an engine for economic growth and human development: a review of the arguments and empirical evidence'. *Human Rights and International Legal Discourse*, *3*, p. 177.

Denisia, V. (2010). 'Foreign direct investment theories: An overview of the main FDI theories'. *European Journal of Interdisciplinary Studies*, (3), 7 pages.

Dunning, J.H. (2002). Trade, location of economic activity and the multinational enterprise: A search for an eclectic approach. In J.H. Dunning (ed.), *Theories and paradigms of international business activity* in The selected essays of John H. Dunning (pp. 52–76). Cheltenham: Edward Elgar Publishing Limited.

Dwyer, L. and Forsyth, P. (1994) 'Motivation and impacts of foreign tourism investment'. *Annals of Tourism Research*, *21*(3), pp. 512–537.

Dwyer, L., Forsyth, P., Madden, J., and Spurr, R. (2000) 'Economic impact of inbound tourism under different assumptions about the macroeconomy', *Current Issues in Tourism*, 3(4), pp. 325–363.

Dwyer L., Forsyth, P. and Spurr, R. (2003) 'Inter-industry effects of tourism growth: some implications for destination managers'. *Tourism Economics*, *9*(2), pp. 117–132.

Dwyer, L. (2015) 'Globalization of tourism: Drivers and outcomes'. *Tourism Recreation Research*, *40*(3), pp. 326–339.

Dwyer, L. and Čavlek, N. (2019) 'Economic globalisation and tourism'. In D. Timothy (ed.) *Handbook of Globalisation and Tourism*. Edward Elgar Publishing, Cheltenham, UK

Elkomy, S., Ingham, H. and Read, R. (2016) 'Economic and political determinants of the effects of FDI on growth in transition and developing countries' *Thunderbird International Business Review*, *58*(4), pp. 347–362.

Endo, K. (2006) 'Foreign direct investment in tourism—flows and volumes'. *Tourism Management*, *27*(4), pp. 600–614.

European Commission (2016) 'Towards a Foreign Direct Investment (FDI) attractiveness scoreboard'. OECD, Directorate General for Internal Market, Industry, Entrepreneurship and SMEs Unit, 1. Economic Analysis, May.

Fang, J., Gozgor, G., Paramati, S.R., and Wu, W. (2020) 'The impact of tourism growth on income inequality: Evidence from developing and developed economies'. *Tourism Economics*, pp. 1–23.

Fauzcl, S., Seetanah, B., and Sannassee, R.V. (2017) Analysing the impact of tourism foreign direct investment on economic growth: Evidence from a small island developing state. *Tourism Economics*, *23*(5), pp. 1042–1055.

Feng, S. (2020) 'Technology transfer spillover from FDI-a comprehensive literature review. In *4th International Symposium on Business Corporation and Development in South-East and South Asia under B&R Initiative (ISBCD 2019)* (pp. 1–11). Atlantis Press, Netherlands.

García, F., Jin, B., and Salomon, R. (2013) 'Does inward foreign direct investment improve the innovative performance of local firms?'. *Research Policy*, *42*(1), pp. 231–244.

Gaur, A.S., Pattnaik, C., Singh, D. and Lee, J.Y. (2019) 'Internalization advantage and subsidiary performance: The role of business group affiliation and host country characteristics'. *Journal of International Business Studies*, *50*(8), pp. 1253–1282.

Ghebrihiwet, N. (2017) Acquisition or direct entry, technology transfer, and FDI policy liberalization. *International Review of Economics & Finance*, *51*, pp. 455–469.

Groh, A.P. and Wich, M. (2012) 'Emerging economies' attraction of foreign direct investment'. *Emerging Markets Review*, *13*(2), pp. 210–229.

Gudowski, J. and Piasecki, R. (2020) 'Foreign direct investment from emerging markets. Theory and practice'. *Comparative Economic Research. Central and Eastern Europe*, *23*(2), pp. 7–19.

Iwasaki, I. and Tokunaga, M. (2016) Technology transfer and spillovers from FDI in transition economies: A meta-analysis. *Journal of Comparative Economics*, *44*(4), pp. 1086–1114.

Jayaraman, T.K., Chen, H. and Bhatt, M. (2014) 'Research note: contribution of foreign direct investment to the tourism sector in Fiji: an empirical study'. *Tourism Economics*, *20*(6), pp. 1357–1362.

Kang, Sung Jin, and Hongshik Lee (2011) 'Foreign direct investment and de-industrialisation.' *The World Economy*, *34*(2), pp. 313–329.

Kudina, A. and Pitelis, C. (2014) 'De-industrialisation, comparative economic performance and FDI inflows in emerging economies'. *International Business Review*, *23*(5), pp. 887–896.

Krugman, P. (2000) Fire-sale FDI. In *Capital flows and the emerging economies: theory, evidence, and controversies* (pp. 43–58). University of Chicago Press, Chicago, USA.

Li, X., Quan, R., Stoian, M.C. and Azar, G. (2018) 'Do MNEs from developed and emerging economies differ in their location choice of FDI? A 36-year review'. *International Business Review*, *27*(5), pp. 1089–1103.

Liu, X., C. Shu, C., and P. Sinclair (2009) 'Trade, foreign direct investment and economic growth in Asian economies' *Applied Economics*, *41*(13), pp. 1603–1612.

McGaughey, S.L., Raimondos, P., and la Cour, L. (2020) 'Foreign influence, control, and indirect ownership: Implications for productivity spillovers'. *Journal of International Business Studies*, pp. 1–22.

Maryam, J. and Mittal, A. (2020) 'Foreign direct investment into BRICS: an empirical analysis'. *Transnational Corporations Review*, *12*(1), pp. 1–9.

Nunnenkamp, P. (2004). 'To what extent can foreign direct investment help achieve international development goals?'. *World Economy*, *27*(5), pp. 657–677.

Nunkoo, R. and Seetanah, B. (2018) 'Foreign direct investment and tourism development: A theoretical and empirical review' Sage handbook of tourism management. C. Cooper (ed.) Chapter 33, pp. 556–568.

Park, B.I. and Xiao, S. (2017) 'What drives outward FDI from emerging economies? The interplay between exploration orientation and home-country institutional changes'. *Canadian Journal of Administrative Sciences/Revue Canadienne Des Sciences de L'administration*, *34*(4), pp. 387–400.

Pathan, S.K. (2017) 'An empirical analysis of the impact of three important aspects of Eclectic Paradigm on Foreign Direct Investment (FDI).' *International Research Journal of Arts & Humanities (IRJAH)*, *45*(45), pp. 35–50.

Perić, J. and Radić, M.N. (2016) Development impact of FDI in hotel: Case study of Terme Tuhelj in Croatia. *Alma tourism-Journal of Tourism, Culture and Territorial Development*, *7*(13), pp. 66–78.

Samimi, A.J., Sadeghi, S., and Sadeghi, S. (2013) 'The relationship between foreign direct investment and tourism development: evidence from developing countries'. *Institutions and Economies*, *5*(2), pp. 59–68.

Selvanathan, S., Selvanathan, E.A. and Viswanathan, B. (2012) 'Causality between foreign direct investment and tourism: empirical evidence from India'. *Tourism Analysis*, 17(1), pp. 91–98.

Sharma, R.R. (2016) 'Taxing and subsidizing foreign investors'. *Economics Faculty Working Papers 50*. Colgate University. http://commons.colgate.edu/econ_facschol/50.

Sheldon P. and Dwyer, L. (2010) 'The global financial crisis and tourism: Perspectives of the academy'. *Journal of Travel Research*, *49*(1), pp. 3–4.

Snyman, J.A. and Saayman, M. (2009) 'Key factors influencing foreign direct investment in the tourism industry in South Africa'. *Tourism Review*. 64(3), pp. 49–58.

Sokhanvar, A. (2019) 'Does foreign direct investment accelerate tourism and economic growth within Europe?' *Tourism Management Perspectives*, *29*, pp. 86–96.

Tamazian, A., Chousa, J., and Vadlamannati, K. (2009) 'Does higher economic and financial development lead to environmental degradation: evidence from BRIC countries. *Energy Policy*, *37*(1), pp. 246–253.

Tiwari, A. (2011) 'Tourism, exports and FDI as a means of growth: evidence from four Asian countries'. *Romanian Economic Journal*, *14*(40), pp. 131–151.

UNCTAD (2008). World Investment Report, TNCs and the Infrastructural Challenge, United Nations, Geneva.

Yazdi, K.S., Nateghian, N., and Sheikh Rezaie, N. (2017) 'The causality relationships between tourism development and foreign direct investment: an empirical study in EU countries'. *Journal of Policy Research in Tourism, Leisure and Events*, *9*(3), pp. 247–263.

3 Local Impacts and Perceptions of Tourism Foreign Direct Investment (TFDI)

A Conceptual Framework

H. Cristina Jönsson and Nicola J. Palmer

Introduction

For many years it was commonly believed that wealth through tourism development would eventually "trickle down" and benefit the local community through multiple channels such as employment, public welfare and family networks. Critics of the neo-liberal paradigm have challenged the magnitude of anticipated multiplier effects (Mosedale, 2011) and questioned the extent to which the development of tourism increases inequality between countries and "between those involved and not involved in tourism within destinations" (Spenceley & Meyer, 2012, p. 300). In the context of tourism, FDI has been acknowledged to be "one of the routes through which developing countries can carry out tourism" yet, "its implications, have been relatively little studied" (UNCTAD, 2007, p. xi).

FDI plays an important role in the global economy and many developing countries consider it to be a fundamental driver of development. Econometric analyses of FDI and economic development are well-established, but their conclusions remain unclear; some analyses indicate a positive impact on local economies while other analyses remain pessimistic. Little is known about the extent of FDI in tourism at large and, more specifically, understanding of local residents' perceptions of its impacts is lacking. There are still inconclusive arguments for and against the role of FDI inflows in enhancing economic growth and human development in any country (Sultanuzzaman, Fan, Akash, Wang, & Shakij, 2018). It remains a matter of debate whether FDI inflows are beneficial or not to socio-economic growth and what governments should do to attract and use FDI inflows effectively.

Tourism-Related Foreign Direct Investment (TFDI)

The importance of FDI is perhaps one of the most significant features of economic globalisation. It has been argued that FDI can have important positive effects on a host country's development effort (see for example, Davidson & Sahli, 2015; Samimi, Sadeghi, & Sadeghi, 2017; Dinh, Vo, & Nguyen, 2019). FDI supplies direct capital financing and can provide a host

DOI: 10.4324/9781003155492-5

country with technology and know-how while encouraging economic linkages with local firms. Furthermore, FDI can generate employment, raise productivity, enhance competitiveness of the domestic economy through transfer of skills and technology, strengthen infrastructure and enhance exports, which can help reinvigorate an economy. Consequently, FDI is viewed as an engine for economic growth and development. With tourism being increasingly important to many economies, at all stages of development, investment is critical for further development of the tourism sector. Tourism is of economic importance as a generator of tax revenues, jobs and increasing tourist expenditure patterns, which may encourage public officials to re-examine the role of tourism as a national economic development generator. In tourism-dependent countries, tourism is often reported to be accepted by governments as a panacea for the country's economic ills (Scheyvens, 1999; Lanfant & Graburn, 2016; Giampiccoli & Mtapuri, 2020). There is a need to shift away from neo-liberalist tenets to consider stewardship of a more inclusive economy. The value in acknowledging broader stakeholder perspectives on how TFDI impacts tourism-dependent countries, such as Barbados, at a local level should not be under-emphasised.

Tourism may be one of the fastest and most effective ways in which small economies can access world markets (Croes, 2006). Due to its distinctiveness in global trade, tourism moves consumers to the product rather than transporting the product to the consumers. The uniqueness of the tourism industry is distinguished in the complex tourism value chain (e.g. consumers/travellers, wholesalers, agents, tour operators, service providers, vendors and natural and cultural attractions). Tourism has been known to generate domestic and foreign investment, foreign exchange earnings, enhance incomes and new jobs and be the creator of inter-sectoral linkages such as agriculture and tourism, for example (Faladeobalade & Dubey, 2014). As a result, many countries are looking to tourism as a potentially promising avenue for economic and social development. Indeed, UNCTAD (2007, 2008) identifies FDI as one of the routes through which developing countries can achieve economic advantage. FDI encompasses capital flows, and is a package of long-term capital, technology and management expertise, and productive capacity that can contribute to development, living conditions and the natural environment of billions of people.

The tourism sector is relatively capital-intensive in terms of infrastructure, knowledge and access to global market and distribution chains, therefore "a country's investment competitiveness goes beyond attracting FDI. It is determined by the country's ability to bring in, retain, and leverage private investment for inclusive and sustainable economic growth" (World Bank, 2017 p. xi). However, FDI is not immune from criticism and it is argued that excessive accommodation of foreign investment risks the phenomenon known as "Dutch Disease" (Botta, Godin, & Missaglia, 2016), i.e., being over-reliant on foreign investment which, in turn, presents a high level of risk to the host country's economic resilience. Many less economically developed countries

(LEDCs) that depend on financial inflows from tourism suffer from lack of competitiveness from other sectors and industries. They also do not generate enough goods and produce internally to sustain their tourism industry. As a result, imports and economic leakages are unavoidable. Although much has been written about leakages to the balance of payments, Xu (2018) states that empirical evidence of the effect of FDI on a country's tourism balance sheet is scarce. The "Dutch Disease" phenomenon can be seen, for example, in the Caribbean where there is an over-reliance on the tourism industry, lack of competitiveness and limited production. It has been argued that this phenomenon can be mitigated through, for instance, local procurement and production initiatives that can stimulate local entrepreneurs to produce new items and sell to other markets (Zhang & Yang, 2019). However, in most island destinations with limited resources, these mitigations often difficult to enact and represent a somewhat unrealistic prospect.

TFDI Impacts

It is argued that the benefits of TFDI outweigh its costs mainly through its impact on tourism flows that lead to net benefits for local residents (Dwyer, Forsyth, & Dwyer, 2020). Beyond promoting growth, TFDI has other potentially desirable features that can assist with resilient development such as, helping to reduce adverse shocks to the poor resulting from financial instability (e.g., as witnessed through the 1997 Asian Crisis). Another attractive feature of TFDI is that it offers improved environmental and labour standards because foreign investors are concerned about their reputation in markets. This can translate to the desirability of high school educational standards and the generation of taxes that help support the development of a safety net for the poor. Moreover, foreign investors may invest substantially into community development. The welfare effects of FDI in tourism have been examined by Tang and Tan (2013). They concluded that FDI generally raises socio-economic welfare gains associated with tourism (e.g., increased tourism expenditure, productivity growth, employment and increase in tourism exports) and, for this reason, FDI should not be restricted or discouraged. Fauzel's (2020) examination of FDI's role in small island economies' tourism development and overall economic growth found that policy encourages FDI inflow, especially into tourism sectors in small island regions. This cumulates in increased overall tourism activity and greater economic development. Additionally, Campbell (2003) used regression analysis and annual data in his study of the impact of FDI on Barbados 1970–1999, finding that FDI had a negative impact on current account (i.e., exports and imports of goods and services) in the short and the long run. Thus, any FDI benefits would be offset by imports and repatriation of profits (i.e., leakage), resulting in pressure on the country's foreign reserves. Despite having a booming tourism sector stimulated by mass foreign investment, many small destinations in the developing world face enormous challenges in overcoming poverty and inequality. It is important to note that the correlation between TFDI and tourism growth does not in itself guarantee local economic growth. In this regard, Brohman (1996) contributes substantially to the literature by clarifying that for the positive correlation which exists between FDI and tourism growth to be effective it must also be substantiated by other intervening variables. These variables include, but are not limited to, management styles and power relationships (Mowforth & Munt, 2015); and existing infrastructure, and economic growth rates (Kishor & Singh, 2015). In the following section, some of these essential independent variables of TFDI and their inter-relationships are discussed.

Micro-Level Impacts

The OECD (Organisation for Economic Development and Cooperation, 2001) has argued that FDI has a particularly substantial impact on income inequality in destinations. However, whereas Trans-national Corporations (TNCs) are often regarded as "exploiting cheap labour" (OECD, 2001: 309), empirical evidence suggests that in fact TNCs have a tendency to pay higher wages than local firms, even inducing local firms to increase their wage scales (Javorcik, 2015). These pay increases perhaps have some bearing on why openness to TFDI by less developed nations is perceived as being high according to Barrowclough (2007), despite a lack of policy implementation to capitalise on those opportunities. There is a need to question whether lack of supportive policies is indicative of a lack of protection for locals. In their study of Vietnam, Haley and Haley (1997) revealed that TFDI had negative impacts such as land price inflation that displaced local communities, and led to environmental degradations which negatively impacted the fishing sector. These observations encouraged the government of Vietnam to subsequently implement regulations regarding the permissible level of FDI specifically into the hotel sector (Haley & Haley, 1997).

Research by Mowforth and Munt (2015) found that given that expatriates usually manage international hotels in less developed countries, this creates a greater potential for foreign dominance and the aggravation of existing international versus local power structures. In spite of this increased foreign presence and resulting expanded power distance, however, UNCTAD (2007)'s study of seven less developed countries (Bhutan, the Dominican Republic, Kenya, Morocco, Sri Lanka, Tunisia and the United Republic of Tanzania) indicated that foreign hotels in least developed and emerging tourism economies had higher staff to guest ratios (8:1) than local properties (1:1 or 1:2). This adds some credence to the notion the TFDI has a substantial role in local employment generation but it brings into question the impact of international management in foreign properties on employees, management styles, and employment conditions.

Fortanier and Van Wijk's (2010) TFDI research in Tanzania, Ethiopia and Mozambique found no evidence to support the predisposition of foreign entities to employ a disproportionate amount of persons vis-à-vis local entities. Their work suggests that with respect to employment, foreign properties have no disproportionately adverse impact on the quantity of persons employed in the sector. The question arises as to whether adverse impacts of TFDI in terms of quantity of employment are inconsistent across destinations, based on socio-economic backgrounds.

Notwithstanding the FDI contribution to the level of employment in the tourism sector, there appears to be a strong and consistent position in the existing literature that TFDI has a more consistent impact on the quality of employee. Fortanier and Van Wijk (2010) found that this was largely driven by higher expatriate to local employee ratios. In essence, hotels with higher expatriate ratios were found to have completed more formal training than those with lesser expatriate ratios and even more so than local properties. Perhaps offering even more validation to this finding is the fact that no evidence was found to support the hypothesis that foreign properties of higher standard offer more formal training than lower quality foreign properties but rather that training across all FDI properties is fairly consistent.

This high quality, formal training is certainly beneficial to the local population and one may argue it even somewhat counters certain aspects of Dunning's (1979) OLI framework which places a heavy emphasis on international firms "exploiting" their unique advantages in local communities. That is not to insinuate though that these international tourism firms do not benefit from that unique advantage. Instead, as epitomised through the consistency FDI properties have in enhancing the quality of employee, this instance avows that this perceived "exploitation" of their unique resources is a basis upon which substantial qualitative investments are made in the human capital. Moreover, interestingly the Williams and Deslandes (2008) study of FDI in Jamaica clearly suggests that FDI in foreign nations is not a monopoly-based competitive strategy. Spanish investors investing in Jamaica were argued to have faced significant competition from other Spanish investors who followed suit. Based on these suggestions it appears therefore to be a reasonable conclusion that when one investor enters the market and raises the quality of the human capital, the competitors who follow suit will also have to make similar investments in raising the quality of human capital in order to be competitive.

Macro-Level Impacts

The Williams and Deslandes (2008) study can be commended for its analysis of other impacts of FDI in Jamaica; however, it begs the question as to whether TFDI has some greater influence in attracting other TFDIs to a particular destination. Certainly if this is so, this would have a significantly positive impact on destinations for the causal relationship it creates between one set of FDI and the further formulation of another distinct set of capital for another investment. In essence, it was argued by Williams and Deslandes

(2008) that TFDI has no correlation to FDI in other sectors (supported by the work of Parys & Sebastian 2010). However, the Williams and Deslandes study certainly supports the validity of any question as to the degree of impact that one set of tourism FDI has on other tourism FDI initiatives. While the injection of capital appears to influence the attraction of additional capital into the sector, Parys and Sebastian (2010) also identify another factor which has a substantial impact on the stimulation of FDI into the sector. They found that post-2003 when the Antiguan government implemented favourable tourism tax incentive policies, tourism investment into the country rose substantially. This reinforces Dunning's (1979) ideas about the influence of location-specific advantages on attracting FDI.

Although tax incentive programmes have positive influences on investment, there are other location-specific advantages which serve to be "pull factors". The study by Singh, McDavid and Birch (2006), for example, alludes to factors such as existing tourism industries, infrastructure, economic growth rates and openness to foreign investment as principal variables in attracting FDI. The authors conclude that the existing size of the market is not a substantial pull factor variable. Support for this suggestion could also be vested in the Williams and Deslandes (2008) study which arguably analyses the formulation of FDI capital from a supply-driven perspective. That is to say, a greater interest is placed on analysing factors which directly impact the supply (such as infrastructure and incentives) than considering whether demand for the presence of foreign entities actually exists. This is perhaps not surprising given that there has been consistent evidence supporting the notion that FDI stimulates tourism growth (Craigwell & Moore, 2007; Roudi, Arasli, & Akadiri, 2019; Samimi et al., 2017). In fact, in Croatia, Perić and Radić (2016) found that more than simply stimulating economic growth, FDI played a substantial role akin to a causal effect on the number of tourist arrivals.

While there appears to be consistency in the literature regarding the impact that FDI has on tourism inflows, as discussed in the Selvanathan, Selvanathan and Viswanathan (2009) report, there is some level of incongruity regarding whether that impact represents a reciprocal relationship. For example, Craigwell and Moore (2007) found tourism to have a causal relationship on FDI in only one third of all cases analysed whereas other researchers (Selvanathan et al., 2009) found no relationship at all. In fact, Williams and Deslandes (2009) found that demand, or the lack thereof, for foreign properties does not necessarily have a substantial impact on the decision to invest capital in establishing a foreign property in a destination.

The study by Haley and Haley (1997) stands as an apt example of the negative occurrences which can force governments to place controls on FDI from a restrictive perspective. Conversely however, Pesakovic and Saunders (2009) also give representation to the consequences of unfettered and uncontrolled FDI growth. They proposed that in the Bahamas, there has emerged a "dual economy" with substantial control in the sector being placed in the hands of foreigners. Pesakovic and Saunders (2009) argue that it is governmental policy failure to ensure sufficient benefits for the local population which has facilitated this economic division. As a result, there exists a foreign economy and a Bahamian economy where the latter is essentially subservient to the former. These studies certainly reinforce the importance of fair participatory governance and the necessity for local stakeholders to have strong political voices.

Given an established strong correlation and causal relationship drawn between FDI and tourism growth, it is not surprising that the UNCTAD (2007) report would consider that FDI into the sector creates additional secondary benefits such as local infrastructural development. The extent to which these impacts are quantified however is very limited. This most likely in large part has to do with the limited amount of research regarding the actual direct and indirect impact that tourism FDI has in developing countries (UNCTAD, 2007). A lack of research is often precipitated in the first instance by the fact that many destinations fail to collect tourism data in systematic ways such as using tourism satellite accounting systems (TSA) (a standard statistical framework and the main internationally-shared tool for the economic measurement of tourism). Notwithstanding these limitations however, there exists a strong recognition in the UNCTAD (2007) study that FDI properties only invest in countries where there is some recognised advantage or benefit either through local competition or through the use of some local resource (such as labour or sun, sea and sand product offerings). This is in essence a repetition of the principles of Dunning's (1979) OLI framework suggesting that foreign properties have some affinity towards some form of "exploitation".

If one accepts the basis of Dunning's (1979) OLI framework which argues that TNCs exist to capitalise on these specific sources of sustained competitive advantage, then it seems logical to question whether there are certain attending consequences which would be more likely to ensue from foreign investment compared to local investment. The study by Sheng and Tsui (2010) supports the argument that FDI properties result in certain spin-off negative impacts which are more specific to the foreign investment than local investment. They found statistically significant correlations between TFDI and negative consequences like leakages, real estate bubbles and the crowding out of local enterprises. However, we should not overlook the formalised training opportunities highlighted earlier by Fortanier and Van Wijk (2010) that indicate potential positive FDI impacts in relation to human capital development. In essence therefore, literature does exist to support the argument that there is some difference in the resulting impact of TFDI compared to local investment and that this is mixed in terms of potential costs and benefits. There remains a need to better isolate and understand some of these impacts from a local perspective by considering local voices to supplement the findings of previously acknowledged studies that have focused and relied on econometric data.

Types of FDI in Tourism

There is a strong presence of non-equity foreign investment in TFDI where the separation of ownership and control often occurs through for example leasing agreements, management contracts and franchise agreements. The tendency of TNCs to enter new markets through non-equity investment rather than through FDI is greater in service industries such as tourism (e.g., hotels, restaurants and car rentals). Various entry mode strategies have been presented by Root (1994), highlighting FDI to be equity-based. Companies that operate mainly through non-equity foreign investments are not captured in FDI data (e.g., stock and flow, economic activities by foreign affiliates) hence the figures of tourism FDI are not a true representation of the extent of investment (FDI and non-equity foreign investment). To understand the true value of tourism FDI it is necessary to include non-equity forms of foreign investment. This makes the examination of the importance of tourism FDI more realistic, since this crucial type of tourism investment opens up markets, facilitates access to markets and creates demand, which leads to greater marketability and attracts visitors to destinations. On the other hand, the

> commonly held view about FDI in tourism in developing countries is that there is too much of it, that it is dominant, and that TNCs do not disperse the benefits of tourism sufficiently widely through the host economy
>
> (UNCTAD, 2007: 7)

Mergers and acquisitions (M & As) are often used to enter a foreign market in the services sector (UNCTAD, 2010) and mainly take place in hotel deals in developed countries. This has to do with the speed and efficiency in purchasing an existing distribution network or trademark (brand) compared to starting a new business. Froot and Stein (1991) examined entry by acquisition and suggest that low values of the host country encourage FDI, arguing that currency depreciations make acquisitions by foreign firms more profitable by creating undervalued assets. Greenfield investment, which is more common for resort development, is likely to be predominant in tourism. This type of investment is where a parent company builds its operations in a foreign country from the ground up. In addition to the construction of new production facilities, these projects can also include the building of new distribution hubs, offices and living quarters. Joint ventures with local partners are often used in the establishment of new tourism facilities (UNCTAD, 2007). TFDI activities generally occur in hotels, restaurants and car rental companies, airlines, tour operators and travel agents. Table 3.1 re-presented from the UNCTAD report (2007) applies Dunning's (1979) OLI approach to these four tourism-related activities and outlines: ownership advantages; locational advantages; internationalisation factors; and organisational forms.

Table 3.1 OLI advantages by tourism industry activity

Industry activity	*Ownership advantages*	*Locational advantages*	*Internalisation factors* *(+) encourages FDI* *(−) encourages other investment modes*	*Organisational forms*
Hotels	Experience in home countries in supplying upmarket services Experience in training key personnel Quality control systems (e.g. management, procurement) Referral system (GDS) Economies of geographical specialisation, access to inputs	Location-bound when selling a foreign service Exports through tourists/business people visiting home or host country	Investment in hotels is capital-intensive (+) Quality control can be ensured through non-equity forms (−) Governments prefer non-equity forms (−) Referral systems are centrally coordinated without equity control (−) Growing brand recognition for new TNCs from the South (+) Lack of managerial expertise in host country (−)	Vary according to positive and negative influences on equity/non-equity decision, because both forms can protect ownership advantages
Restaurants and car rentals	Brand name and image Reputation and experience Referral system (GDS) Economies of scale and scope Tie-up deals with airlines and hotels	Location-bound Foreign earnings through tourists and business people visiting exporting countries	Franchising can protect quality (−)	As with hotels, forms vary because ownership advantages can be protected by contract

(*Continued*)

Industry activity	*Ownership advantages*	*Locational advantages*	*Internalisation factors (+) encourages FDI (−) encourages other investment modes*	*Organisational forms*
Airlines	Highly capital-intensive	Logistical management	Role is essentially location-linking (−)	International services do not require FDI
	Government support measures and/ or control over routes of foreign carriers	Advantages of vertical integration	Need for local sales office, access to terminal and maintenance and support facilities (+/−)	Growth in alliances and affiliations (e.g. code-sharing)
		Quality control		
			Growth in alliances and code-sharing (−)	
			Liberalisation of markets (+)	
Tour operators/ travel agents	Reputation of providing satisfactory experience	Need for local tour agents and support facilities	Coordination of itineraries, packaging of services, need for quality control of ancillary services for tourists (−)	FDI is rare; mostly firms have only local agents
	Economies of scope (travel portfolio offered)	Customers initially originate from home country?	Economies of transaction costs from vertical integration (+)	
	Bargaining power	Costs of supplying local facilities usually lower	Growth in e-commerce and increasing role of local tour operators (–)	
	Quality of deals made with airlines, hotels, cruise companies and other associated services	Fiscal incentives and infrastructure facilities		

Source: After UNCTAD (2007).

This table demonstrates internationalisation factors that encourage FDI and factors that encourage other modes of investment.

TFDI Trends

UNCTAD (2010) highlighted three general trends in TFDI. Firstly, it was proposed that although tourism is a substantially large industry, it appears to be one of the least globalised with low levels of TFDI evident in both developed and developing countries when compared to other industries. As an example, the UNCTAD (2010) report highlighted that TFDI outflows from the UK were just $34,404 million or 2.5% of that nation's total FDI outflows. Secondly, the report proposed that TFDI 85–90% of tourism TNCs was located in developed countries. Moreover, to further exacerbate this finding, it is also widely accepted and acknowledged in the UNCTAD (2010) report that TFDI is often most associated with four specific types of tourism activity (see Table 3.2). Thirdly, there was a proposal within the report (UNCTAD, 2010) that notwithstanding these prior two aforementioned trends, TFDI inflows to developing nations were increasing drastically.

The shift in the concentration of TFDI can be better understood through a review of the study by Zhang and Jensen (2007) which *inter alia* provides a reasonable explanation as to how FDI which is related to tourism impacts the proliferation of further FDI into the sector. Zhang and Jensen (2007) argued that tourism FDI in developing and lesser developed nations tended to be directed more towards facilitating the development of local tourism

Table 3.2 Frequency occurrence of TFDI by activity

Activity	*Frequency with which TFDI occurs within activity*		
	Most frequent	*Occasional*	*Rare*
Hotels and similar	✓		
Restaurants and similar	✓		
Second homes	✓		
Passenger transport and rental equipment	✓	✓	
Railway passenger transport services		✓	
Air passenger transport services			✓
Road passenger transport services			✓
Water passenger transport services			✓
Passenger transport supporting services			✓
Travel agencies and similar			✓
Cultural services			✓
Sports and other recreational services			✓

Source: After UNCTAD (2010).

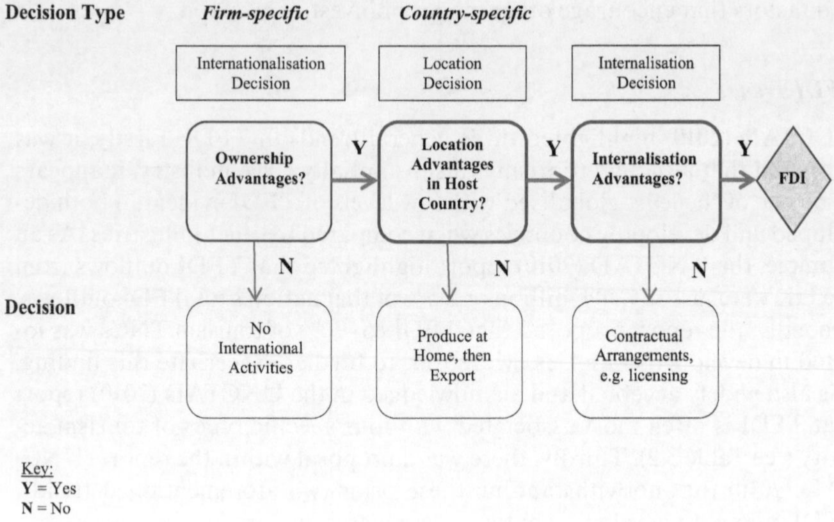

Figure 3.1 Dunning's eclectic approach – OLI.

sectors through investments in the development of infrastructure and technology rather than directly focused on attracting tourists. They explained this revelation by contrasting FDI in OECD economies compared to lesser developing economies suggesting that "there is already excess capacity in the OECD countries, hence investing in additional infrastructure will not in itself help to attract more tourists" (Zhang & Jensen, 2007: 240).

The fact that Zhang and Jensen (2007) argue that the over-capacity is a dominant factor motivating tourism TNCs away from the OECD countries is largely substantiated by the location-specific advantages segment of Dunning's (1979) OLI framework (Figure 3.1). The Zhang and Jensen (2007) study in its pronouncement on over-capacity, when examined in the context of Dunning's (1979) OLI framework assists in explaining how this shift in the concentration of TFDI is being perpetuated. As mentioned by Dunning (1979), these location-specific advantages determine country selection; therefore, it logically follows that the ownership advantages which these TNCs have can be better exploited in locations where that ownership can serve to be of greater distinct competence than in a location where the market has reached a state of relative maturity.

Many developing nations often use tourism development as a means of social and economic development. However, a major obstacle for tourism development is shortage of capital, and many countries look to foreign investors with capital that will develop their tourism industries. Traditionally, the largest tourism FDI source countries have been USA, UK and France (UNCTAD, 2007). Surprisingly, however, this re-concentration of outward

FDI from developed to developing nations has also been accompanied by a rise in FDI to the developing world from other developing economies. Considered a new trend in outward FDI, it is the rise of South–South investment, where a number of TNCs are from developing and transition economies such as the BRIC countries (Brazil, Russia, India and China) (Rangel, 2012).

South–South investment is considered to be proliferated largely as a result of closer cultural, geographic and trade relationships between host and receiving nations. It is viewed as a necessary complement to the more traditional North-South (developed to developing world) investment since it allows developing countries the ability to access capital from other developing nations, especially in instances whereby TNCs in the developed world adopt risk averse policies (Meyer, 2011).

Notwithstanding these emerging trends and re-concentrations in TFDI, there is wide acceptance that TFDI in developing nations has been insignificant. It is only logical that some attempt be made to elucidate the impacts of TFDI in developing regions as a result of this influx of more TFDI to the developing world.

Local Voices and Perceptions of Tourism Impacts

Knowledge is always mediated by pre-existing ideas and values, whether this is acknowledged by researchers or not and human communities in practice have created reasonably firm grounds on which credibility can be judged, whether or not these grounds can be supported in some ultimate sense by means of philosophical reasoning. Perhaps the difficulty in this primarily derives from the inherently fallible and imperfect nature of our understanding of the world in which we live. For this reason, there are no absolute truths, and therefore, one could only speak of contingent validation of social ideas and norms of appropriateness. Truths are proven and disproved every day. Often based on fact, truths are the humanistic extension of facts. Truths take a fact and combine the totality of the environment, situation and unverifiable subjectivity of the individuals or things involved. Truth can be seen as a fact summarised by life, a life is different for everyone, and therefore the truth of a fact can be unique for each person. Facts can change over time while truths and perceptions can change by the second.

Resident perceptions have been widely studied in tourism because of the direct impact of the industry and its related activities on local people. Through both endorsement and resistance, residents have an impact on how well tourism in a destination flourishes (Deery, Jago, & Fredline, 2012). As a result, it is imperative for their voices to be heard. Perceptions have varying meanings and definitions, which range from the simple dictionary definition, "a belief or opinion, often held by many people and based on how things seem" (Cambridge University Press, 2016) to the definition by Pickens (2005: 52) who states that, "a perception is the process by which organisms interpret and organise sensation to produce a meaningful experience of the world". This has been further expanded to include recognition, and the fact that the process of actually receiving stimuli through senses is the first step in developing perception. Perception is the process by which we interpret the worlds around us, forming a mental representation of the environment. The stimuli are interpreted into something meaningful to that person due to past experiences. However, Pickens adds that what an individual interprets or perceives may be substantially different from reality. In his view, everyone lives in their own world therefore there is no uniform consistent reality. Reality is individual and as a result there are many realities, all made up of objective realities (what actually happens) and subjective realities (how our brains make sense of what happens). Reality may be regarded as a subjective concept; therefore, it cannot be assumed that everything is the same for all individuals. Since there is no uniform consistent reality, it is important to take various perceptions into consideration. It has been acknowledged that although reality is open to interpretation at an individual level, there are aspects that may be socially shared. Thus, areas of congruence and dissonance between the realities of individuals may be identified. This is important to note if research is to inform policy responses.

Hence, it is crucial to focus on local residents experiences of TFDI (with and without a connection to FDI companies) but also consider the perspective of the FDI companies in the destination, such as resort hotels, airlines, restaurants and car rental companies. This will allow for a more rounded and meaningful analysis of local TFDI impacts and will offer understanding of local resident perceptions, in particular, which adds to existing TFDI research and provides valuable knowledge for tourism decision makers.

There has been extensive research focusing on community perceptions of tourism and its impacts (e.g. Sinclair-Maragh, Gursoy, & Vieregge, 2015; Kim, Uysal, & Sirgy, 2013; Almeida-García, Peláez-Fernández, Balbuena-Vázquez, & Cortés-Macias, 2016; Stylidis, Biran, & Szivas 2014), although this has included little evaluation of local community perceptions of TFDI. Despite studies on community perceptions of tourism, few researchers have explicitly examined perceptions of local residents toward TFDI generally and, more specifically, in a tourist dependent island destination (e.g., Wortman, Donaldson, & van Westen, 2016), the Caribbean and small island developing states (SIDS) (e.g., Fauzel, Seetanah, & Sannassee, 2016).

Understanding resident perceptions of tourism and its impacts is important for the success of tourism destinations and has therefore been widely examined. However, impacts of FDI in tourism-dependent destinations have been mainly studied from an econometric perspective (e.g., Craigwell & Moore, 2007). Adding a local voice to the existing knowledge in the form of qualitative perceptions will assist in painting a fuller picture of TFDI and its local destination impacts.

There are various factors that can affect or alter residents' perceptions of tourism. Specific influences have been examined in the tourism literature, such as socio-demographic, economic and spatial factors, as well as involvement, and the individual's knowledge of tourism. Socio-demographic factors including age, educational level, gender, community attachment, ethnicity, occupation and income have been the focus of much discourse, with contradicting or conflicting reports pertaining to each specific variable. Some studies (e.g., Tomljenovic & Faulkner, 2000) show contrasting results with regards to age and the perceived impacts of tourism. On the other hand, researchers such as (Teye, Sirakaya, & Sönmez, 2002) have argued there to be a significant relationship between level of education and the perception of tourism; persons with higher levels of education were found to be taking a greater interest in the impacts of tourism and its contribution to improvement in quality of life.

Community attachment and length of residency in the community is another factor that has been argued to influence perceptions of tourism. While some research has been inconclusive with regards to attachment and perceived impacts (Gursoy, Jurowski, & Uysal, 2002), the findings of other studies emphasise the observation that the more attached a resident is to their community, the greater their concern about tourism-associated impacts. It is claimed that residents usually have a sense of community spirit and pride and therefore are more sensitive to developments or projects that affect the community, although the extent of collective interest is open to debate in differing geographical and social contexts (Palmer & Chuamuangphan, 2017).

There is a plethora of research on the economic impacts of tourism that demonstrates that significant economic benefits that can be gained through tourism such as employment generation (Kalaiya & Kumar, 2015), contribution to government revenues, foreign exchange earnings, stimulation of infrastructure development (Webster & Ivanov, 2014) and contribution to local economies (Pratt, 2015). On the other hand, there are however reports of negative economic consequences of tourism development such as export leakages (Rylance & Spenceley, 2017), increase in cost of living (Sinclair-Maragh et al., 2015), national economic dependency on tourism (Chaperon & Bramwell, 2013) and the economic impacts of the seasonality of tourism (Martín, Aguilera, & Moreno, 2014).

Spatial factors such as distance of tourism development to residential area, and the geographical location of the tourism area and size have also been studied. Some research has shown that living within close proximity to tourism areas can lead to residents having increased negative views on its impacts (Harrill & Potts, 2003). In contrast, other studies have suggested that the further the distance between residence and tourism development, the less positively perceived is the impact (e.g., Teye, Sirakaya, & Sönmez, 2002). It is perhaps self-explanatory that with tourism areas located within residential environs or in close proximity, issues such as increased stress on public transport and other local infrastructures, congestions and increased pressure on local services can arise (Kim, Uysal, & Sirgy, 2013). These occurrences can cause issues for those in close proximity to tourism sites and therefore residents that are furthest away are the ones usually expected to be most in support of tourism development. However, this remains open to debate; while most research to date supports this claim, Jurowski and Gursoy (2004) maintain that residents who live close to tourist development support tourism more than those who live far away.

Residents' ability to voice their opinions and be a part of tourism decision-making processes can have an impact on their perception of tourism. Research shows that when residents are involved with various community activities, they appear to be more favourably disposed towards community change and development (Sharpley, 2014). However, resident involvement is not always possible, particularly in developing countries where tourism projects are often externally initiated.

When it comes to individuals' knowledge of tourism and their level of contact with tourists, Lepp (2006) suggest that residents who are more involved with tourism and directly interacting with tourists are more positively inclined toward the activity and possess favourable perceptions. Lepp (2006) further adds that in locations where residents have little or no prior knowledge of tourism, the activity can be received with distrust, fear and anxiety. Residents that have a higher knowledge of tourism are expected to understand its costs, benefits and impacts and, linked to this notion that familiarity breeds favourability, it has been claimed that local residents working in tourism have been shown to have more favourable perceptions (Brunt and Courtney, 1999).

There have been various approaches to studying perceptions of tourism, its development and its impacts. Each approach offers a different level of insight into residents' perceptions towards tourism. Perceptions based on place attachment, for example, have been studied using Attribution Theory (e.g., Gross & Brown, 2006) while Dependency Theory has been applied to resident perceptions based on livelihood and income reliance. With respect to the latter approach however, Preister (1989) argues that the macro-level orientation of dependency theory limits its application at the local community level and he recognises that it is unable to account for both positive and negative impacts. One of the most frequently used frameworks used to investigate resident perceptions of tourism and its development, in a range of contexts and for differing purposes, is Social Exchange Theory (SET). This is one of the few theories that has been argued to be appropriate to explain both positive and negative perceptions, and suitable for studying the relationship between residents and tourism at the individual and the collective level.

Social Exchange Theory (SET)

Social Exchange Theory (SET), proposed by Thibaut and Kelley (1959), is established as the most accepted framework used to explain residents' reactions to tourism development. SET captures different views based on experiential and psychological outcomes (Nunkoo & Ramkissoon, 2011; Prayag, Hosany, Nunkoo, & Alders, 2013) and it takes social interactions into consideration as an "exchange of resources". SET suggests that individuals are more likely to engage in an exchange when there are expected benefits, however without encountering unacceptable costs (Diedrich & Garcia-Buades, 2009). As such, satisfaction with an exchange is obtained by the evaluation of the outcomes by those involved in relationships of exchange, and the outcomes can be broadly categorised as "economic", "socio-cultural" and "environmental" (e.g. Nunkoo & Ramkissoon, 2012; Andersson & Lundberg, 2013). In addition to SET, the Social Disruption Theory can be used to explain the relationship between local perceptions and tourism development. According to Diedrich and Garcia-Buades (2009), Social Disruption Theory is closely linked to "boomtowns", which in sociology refers to locations that experience a sudden rapid growth in demand for social services and community infrastructure. This leads to a shock (or disruption) for local residents. Once the initial shock wears off, it is claimed that residents adjust to the changes, and with that, the perception of costs decreases. It is when levels of negative impact surpass levels of positive impact that resident perceptions arguably become a cause for concern; negative perceptions can hinder the success of tourism destinations. Consequently, for tourism development and its operation to be successful, it is imperative to take resident perceptions of tourism impacts into consideration.

The Economic Dependence Factor postulates that communities and individuals that are economically dependent on tourism are more favourable to its development (Vargas-Sánchez, Plaza-Mejía, de los, & Porras-Bueno, 2009: 31). Nevertheless, McGehee and Andereck (2004) argue that residents in tourism-dependent communities would rather have less tourism development and its impacts are perceived as negative in comparison to communities that are less economically dependent on tourism. The perception of "costs" and "benefits" can be a significant factor in determining resident perceptions of tourism. Factors that may affect perceptions include: concerns that residents have for their community; the degree to which they are environmentally sensitive; and the extent to which residents use the same resource base as tourists. It should be recognised that these ideas build on the works of earlier researchers and provide a theoretical basis for SET, which assumes that individuals select exchanges after an evaluation of "rewards" and "costs" (Jönsson & Devonish, 2007). Consequently, residents who evaluate the exchange as "beneficial" perceive a positive impact, whereas a negative impact will be perceived by someone who evaluates the exchange as harmful or "detrimental".

Conceptual Framework Based on Academic Literature

Reflecting on the review of literature in this chapter, a conceptual framework of stakeholder impact relations is presented (Figure 3.2). This is based on common discourses within the existing literature on TFDI that envisage

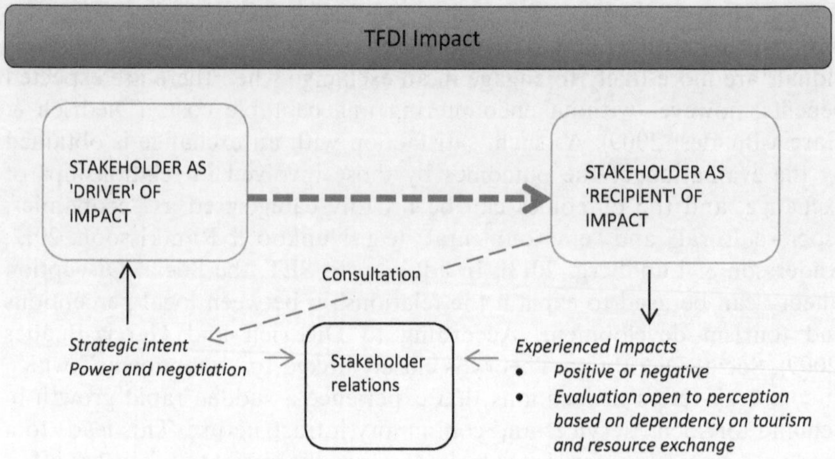

Figure 3.2 Conceptual framework of stakeholder: impact relations based on existing research on impacts of TFDI.

impacts as malleable, open to being directed by stakeholders who act as "impact drivers". Some semblance of agency is also suggested by a shared notion that "impact recipients" are also able to evaluate impacts differently based on levels of dependency on tourism and resource exchange strategies. Here, the notion of capability and capacity building arises; if community stakeholders, in particular, do not possess the capabilities to harness, or even recognise, opportunities then the extent to which recipients might influence impacts received is questionable. One deficit in the existing body of literature around TFDI impacts is a lack of attention paid to power, influence and local voices. This is exacerbated by the exclusive adoption of quantitative methods, an emphasis on macro-level impacts in contrast to micro-level impacts and a quest to capture factual, statistical information as with value chain analysis and UNCTAD's approach. Figure 3.2 summarising existing literature on TFDI impacts also largely separates the "driver" of impacts from the "recipients" of impacts; consultation and opportunity to influence the stakeholder: impact relationship.

Conclusion

This chapter has presented a critical review of existing literature pertaining to local impacts and resident perceptions of TFDI in tourism-dependent destinations. It has identified a gap in the literature in terms of TFDI and the intersection of economic theories and the perceptions of it by local communities who are affected. It has furthermore laid the theoretical foundations for future research in terms of both economic and social theories related to TFDI and those relating to the attitudes of local communities towards it, presenting a conceptual framework of stakeholder impact relations. Dunning's (1979) OLI framework can be applied to future research in order to examine the locational advantages in a case study location. SET allows for the capturing of different perceptions of rewards and costs between individuals and has been often applied in tourism research. Therefore, the ideas brought together here may be useful in developing an overarching framework or lens to examine local perceptions of TFDI in a case study location, offering an opportunity to shift away from traditional neo-liberalist tenets in the study of FDI and tourism, to promote consideration of stewardship of more inclusive economic and social development. This will contribute to the literature by enabling examining local perceptions of TFDI. In bringing FDI and local perceptions of tourism literature together, it is hoped that the discussion presented provides a useful approach to better understand the impacts of TFDI on local people in a tourism-dependent, small island economy and potentially in other lesser developed economies where local resident voices have often been overlooked.

References

Almeida-García, F., Peláez-Fernández, M. Á., Balbuena-Vázquez, A., & Cortés-Macias, R. (2016). Residents' perceptions of tourism development in Benalmádena (Spain). *Tourism Management*, *54*, 259–274.

Andersson, T.D. & Lundberg, E. (2013). Commensurability and sustainability: Triple impact assessments of a tourism event. *Tourism Management*, *37*(1), 99–109.

Barrowclough, D. (2007). Foreign investment in tourism and small island developing states. *Tourism Economics*, *13*(4), 615–638.

Botta, A., Godin, A., & Missaglia, M. (2016). Finance, foreign (direct) investment and Dutch disease: The case of Colombia. *Economia Politica*, *33*(2), 265–289.

Brohman, J. (1996). New directions for tourism in the Third World. *Annals of Tourism Research*, *9*(3), 331–358.

Brunt, P. & Courtney, P. (1999). Host perceptions of sociocultural impacts. *Annals of Tourism Research*, *26*(3), 493–515.

Cambridge University Press (2016). *Cambridge online dictionary*, Cambridge Dictionary online. Retrieved from https://dictionary.cambridge.org/.

Campbell, T. (2003). Foreign direct investment inflows and their impact on Barbados' current account: Implications for policy, facing globalization: Impact and challenges for Barbados and the Caribbean 2002. *Central Bank of Barbados*, 5–24.

Chaperon, S. & Bramwell, B. (2013). Dependency and agency in peripheral tourism development. *Annals of Tourism Research*, *40*, 132–154.

Craigwell, R. & Moore, W. (2007). Foreign direct investment and tourism in SIDS: Evidence from panel causality tests. *Tourism Analysis*, *13*(4), 427–432.

Croes, R.R. (2006). A paradigm shift to a new strategy for small island economies: Embracing demand side economics for value enhancement and long term economic stability. *Tourism Management*, *27*(3), 453–465.

Davidson, L. & Sahli, M. (2015). Foreign direct investment in tourism, poverty alleviation, and sustainable development: A review of the Gambian hotel sector. *Journal of Sustainable Tourism*, *23*(2), 167–187.

Deery, M., Jago, L., & Fredline, L. (2012). Rethinking social impacts of tourism research: A new research agenda. *Tourism Management*, *33*(1), 64–73.

Diedrich, A. & Garcia-Buades, E. (2009). Local perceptions of tourism as indicators of destination decline. *Tourism Management*, *30*(4), 512–521.

Dinh, T.T.H., Vo, D.H., & Nguyen, T.C. (2019). Foreign direct investment and economic growth in the short run and long run: Empirical evidence from developing countries. *Journal of Risk and Financial Management*, *12*(4), 176.

Dunning, J.H. (1979). Explaining changing patterns of international production: In defense of the eclectic theory. *Oxford Bulletin of Economics and Statistics*, *41*, 269–295.

Dwyer, L., Forsyth, P. & Dwyer, W. (2020). *Tourism Economics and Policy*. Bristol, Channel View Publications.

FaladeObalade, T.A. & Dubey, S. (2014). Managing tourism as a source of revenue and foreign direct investment inflow in a developing country: The Jordanian experience. *International Journal of Academic Research in Economics and Management Sciences*, *3*(3), 16–42.

Fauzel, S. (2020). FDI and tourism futures: A dynamic investigation for a panel of small island economies. *Journal of Tourism Futures* *7*(1), 98–110.

Fauzel, S., Seetanah, B., & Sannassee, R. (2016). Analysing the impact of tourism foreign direct investment on economic growth: Evidence from a small island developing state. *Tourism Economics*, *23*(5), 1042–1055.

Fortanier, F. & Van Wijk J. (2010). Sustainable tourism industry development in sub-Saharan Africa: Consequences of foreign hotels for local employment. *International Business Review*, *19*, 191–205.

Froot, K. & Stein, J. (1991). Exchange rates and foreign direct investment: An imperfect capital markets approach. *The Quarterly Journal of Economics*, *106*, 1191–217.

Giampiccoli, A. & Mtapuri, O. (2020). Introducing the investment redistributive incentive model (IRIM): A new redistribution perspective in tourism investment and beyond. *African Journal of Hospitality, Tourism and Leisure*, *9*(1), a58.

Gross, M.J. & Brown, G. (2006). Tourism experiences in a lifestyle destination setting: The roles of involvement and place attachment. *Journal of Business Research*, *59*, 696–700.

Gursoy, D., Jurowski, C., & Uysal, M. (2002). Resident attitudes – A structural modeling Approach. *Annals of Tourism Research*, *29*(1), 79–105.

Haley, U. & Haley, G. (1997). When the tourists flew in: Strategic implications for foreign direct investment in Vietnam's tourism industry. *Management Decision*, *35*(8), 595–604.

Harrill, R. & Potts, D.T. (2003). Tourism planning in historic districts: Attitudes toward tourism development in Charleston. *Journal of the American Planning Association*, *69*, 233–244.

Ivanov, S. & Webster, C. (2014). Transforming competitiveness into economic benefits: Does tourism stimulate economic growth in more competitive destinations? *Tourism Management*, *40*, 137–140.

Javorcik, B.S. (2015). Does FDI bring good jobs to host countries? *The World Bank Research Observer*, *30*(1), 74–94.

Jönsson, C. & Devonish, D. (2007). Preparing for cricket world cup 2007 – Resident's attitudes to the bed & breakfast initiative in Barbados. *International Journal of Event Management IJEMR*, *3*(1), 1–14.

Jurowski, C. & Gursoy, D. (2004). Distance effects on residents' attitudes toward tourism. *Annals of Tourism Research*, *31*, 296–312.

Kalaiya, A.B. & Kumar, A. (2015). Tourism as a development tool: A study on role of tourism in economic development, employment generation and poverty reduction: Special focus on Kachchh. *International Journal of Advance Research in Computer Science and Management Studies*, *3*(7), 189–197.

Kim, K., Uysal, M., & Sirgy, M. J. (2013). How does tourism in a community impact the quality of life of community residents? *Tourism Management*, *36*, 527–540.

Kishor, N. & Singh, R.P. (2015). Determinants of FDI and its impact on BRICS countries: A panel data approach. *Transnational Corporations Review*, *7*(3), 269–278.

Lanfant, M.F. & Graburn, N.H. (2016). International tourism reconsidered: The principle of the alternative. In V.L. Smith & W.R. Eadington, W.R. *Tourism Alternatives* (pp. 88–112). Philadelphia: University of Pennsylvania Press.

Lepp, A. (2006). Residents' attitudes toward Tourism in Bigodi Village, Uganda. *Tourism Management*, *28*(3), 876–885.

Martín, J.M.M., Aguilera, J.D.D.J., & Moreno, V.M. (2014). Impacts of seasonality on environmental sustainability in the tourism sector based on destination type: An application to Spain's Andalusia region. *Tourism Economics*, *20*(1), 123–142.

McGehee, N.G. & Andereck, K.L. (2004). Factors predicting rural residents' support of tourism. *Journal of Travel Research*, *43*, 131–140.

Meyer, D. (2011). Changing power relations. In J. Mosedale (Ed.). *Political Economy and Tourism: A Critical Perspective*. Oxon: Routledge Page 157- 174.

Mosedale, J. (2011). Re-Introducing tourism to political economy. In J. Mosedale (Ed.). *Political Economy of Tourism: A Critical Perspective* (pp. 1–14). Abingdon: Routledge. Mowforth, M. & Munt, I. (2015). *Tourism and Sustainability: Development, Globalisation and New Tourism in the Third World*. London: Routledge.

Nunkoo, R. & Ramkissoon, H. (2011). Developing a community support model for tourism. *Annals of Tourism Research*, *38*(3), 964–988.

OECD. (2001). *Global Forum on International Investment: New Horizons for Foreign Direct Investment*. Paris: OECD.

Palmer, N.J. & Chuamuangphan, N. (2017). Societal values and local responses to appropriate livelihoods, fairness and decision-making involvement in ecotourism: Chiang Rai, Thailand, *Asian Journal of Tourism Research*, *2*(1), 118–149.

Parys, S. & Sebastian, J. (2010). *The Effectiveness of Tax Incentives in Attracting FDI: Evidence from the Tourism Sector in the Caribbean*. Gent: Universiteit Gent.

Perić, J. & Radić, M.N. (2016). FDI-led tourism growth hypothesis: Empirical evidence from Croatian tourism. *European Journal of Tourism, Hospitality and Recreation*, *7*(3), 168–175.

Pesakovic, G. & Saunders, O. (2009). Negotiated foreign direct investment: A case study of the Bahamas. Paper presented at The Sixth International Critical Management Conference, July 13–15, 2009. UK: Warwick Business School, The University of Warwick.

Pickens, J. (2005). *Organizational Behavior in Health Care*. Sudbury: Jones and Bartlett Publishers.

Pratt, S. (2015). The economic impact of tourism in SIDS. *Annals of Tourism Research, 52*, 148–160.

Prayag, G., Hosany, S., Nunkoo, R., & Alders, T. (2013). London residents' support for the 2012 Olympic Games: The mediating effect of overall attitude. *Tourism Management, 36*, 629–640.

Preister, K. (1989). The theory and management of tourism impacts. *Tourism Recreation Research, 14*(1), 15–22.

Rangel, M. (2012). South-South foreign direct investment flows: Wishful thinking or reality. *The Future of South–South Economic Relations* (pp. 111–131). London: Zed Books.

Root, F.R. (1994). *Entry Strategies for International Markets.* New York: Lexington Books.

Roudi, S., Arasli, H. & Akadiri, S.S. (2019). New insights into an old issue–examining the influence of tourism on economic growth: Evidence from selected small island developing states. *Current Issues in Tourism, 22*(11), 1280–1300.

Rylance, A. & Spenceley, A. (2017). Reducing economic leakages from tourism: Value chain assessment of the tourism industry in Kasane, Botswana. *Development Southern Africa, 24*, 295–313.

Samimi, A.J., Sadeghi, S. & Sadeghi, S. (2017). The relationship between foreign direct investment and tourism development: Evidence from developing countries. *Institutions and Economies, 5*(2), 59–68.

Scheyvens, R. (1999). Ecotourism and the empowerment of local communities. *Tourism Management, 20*(2), 245–249.

Selvanathan, S., Selvanathan E.A., & Viswanathan, B. (2012). Causality between Foreign Direct Investment and Tourism: Empirical Evidence from Indi. *Tourism Analysis, 17*(1), 91–98.

Sharpley, R. (2014). Host perceptions of tourism: A review of the research. *Tourism Management, 42*, 37–49.

Sheng, L. & Tsui, Y. (2010). Foreign investment in tourism: The case of Macao as a small tourism economy. *Tourism Geographies, 12*(2), 173–191.

Sinclair-Maragh, G., Gursoy, D., & Vieregge, M. (2015). Residents′ perceptions toward tourism development: A factor-cluster approach. *Journal of Destination Marketing & Management, 4*(1), 36–45.

Singh, D., McDavid, M., & Birch, A. (2006). The determinants of FDI in small developing nation states: an exploratory study. *Social and Economic Studies, 57*(3/4), 79–104.

Spenceley, A. & Meyer, D. (2012). Tourism and poverty reduction: Theory and practice in less economically developed countries. *Journal of Sustainable Tourism, 20*(3), 297–317.

Stylidis, D., Biran, A., Sit, J., & Szivas, E. M. (2014). Residents' support for tourism development: The role of residents' place image and perceived tourism impacts. *Tourism Management, 45*, 260–274.

Sultanuzzaman, M.R., Fan, H., Akash, M., Wang, B., & Shakij, U.S.M. (2018). The role of FDI inflows and export on economic growth in Sri Lanka: An ARDL approach. *Cogent Economics & Finance, 6*(1), 1518116.

Teye, V., Sirakaya-Turk, E., & Sönmez, S. (2002). Residents' attitudes toward tourism development. *Annals of Tourism Research, 29*, 668–688.

Thibaut, J.W. & Kelley, H.H. (1959). *The Social Psychology of Groups.* New York: Wiley. Incur.

Tomljenovic, R. & Faulkner, B. (2000). Tourism and older residents in a Sunbelt Resort. *Annals of Tourism Research*, *27*, 93–114.

UNCTAD, 2007. *FDI in Tourism: The Development Dimension*. New York and Geneva: United Nations.

UNCTAD, 2008. *World Investment Report: Transnational Corporations, and the Infrastructure Challenge*. New York and Geneva: United Nations.

UNCTAD, 2010. *World Investment Report 2010: Investing in a Low-Carbon Economy*, Global. New York and Geneva: United Nations.

Vargas-Sánchez, A., Plaza-Mejía, M., de los Á., & Porras-Bueno, N. (2009). Understanding residents' attitudes toward the development of industrial tourism in a former mining community. *Journal of Travel Research*, *47*(3), 373–387.

Williams, D. & Deslandes, D. (2008). Motivation for service sector foreign direct investments in emerging economies: Insights from the tourism industry in Jamaica. *The Round Table: The Commonwealth Journal of International Affairs*, *97*(396), 419–437.

World Bank (2017). *Global Investment Competitiveness Report 2017/2018: Foreign Investor Perspectives and Policy Implications*. Washington, DC: World Bank.

Wortman, T., Donaldson, R., & van Westen, A. (2016). 'They are stealing my island': Residents' opinions on foreign investment in the residential tourism industry in Tamarin, Mauritius: Opinions on residential tourism in Mauritius. *Singapore Journal of Tropical Geography*, *37*, 139–157.

Xu, C. (2018). The impact of tourism foreign direct investment on economic development: an illustration by development of a tourism satellite account for Wales (Doctoral dissertation, Cardiff University).

Zhang, H. & Yang, Y. (2019). Prescribing for the tourism-induced Dutch disease: A DSGE analysis of subsidy policies. *Tourism Economics*, *25*(6), 942–963.

Zhang, J. & Jensen, C. (2007). Comparative advantage: Explaining tourism flows. *Annals of Tourism Research*, *34*(1), 223–243.

4 Tourism FDI and Sustainable Development in Mauritius. A Dynamic Investigation

Sheereen Fauzel

Introduction

Tourism plays a major role in the economy of developing countries. Small island economies also depend a lot on the tourism sector for their development. This industry contributes to economic growth through various avenues. These can be in terms of both direct and indirect employment creation (Fauzel, 2016), improvement in infrastructure (Briedenhann and Wickens, 2004; Becker and George, 2011; Liasidou, 2013), and even an increase in both local and foreign investment (Fauzel et al., 2015). The international travel and tourism industry represent a major part of the world economy (Lew, 2011). Moreover, this sector generates massive export revenue for various countries (UNWTO, 2001). Destinations having rich natural and cultural heritage exploit this sector to its fullest in order to generate foreign exchange, create employment, and boost government taxes (Hindley and Smith, 1984; Mihalič, 2002).

In 2019, the travel and tourism sector contributed US$8.9 trillion to the world's GDP, which represents 10.3% of global GDP, generated 330 million jobs, and also generated US$1.7 trillion of visitor exports (6.8% of total exports, 28.3% of global services exports). In addition to that, US$948 billion was invested in the sector (Capital Investment).[1] Mauritius is a small island and the tourism sector is an important pillar contributing to its development. In 2019, the tourism industry contributed 7.1% to GDP. More so, there has been massive foreign investment in the tourism sector. While tourism FDI can lead to significant growth in the country, it can as well be harmful for the environment. Hence, sustainable investment is crucial. Therefore, tourism FDI should conform to high environment, social, and governance standards in order to attain sustainability in economic, social, and environmental terms, representing the three essential pillars of development (Narula, 2012).

In Mauritius, FDI inflows have been impressive, mainly increasing between the periods 2001 and 2012. Comparing the period 2001–2006 and 2007–2012, FDI per capita increased by ten-fold while FDI per $1,000 of

DOI: 10.4324/9781003155492-6

GDP and as a percentage of gross fixed capital formation increased by more than five-fold. However, from 2013 to 2016, FDI indicators have been declining. Consider the diagram below showing net FDI inflow trend for Mauritius from 1990 to 2019 (Figure 4.1).

In terms of sectoral distribution of FDI, between the years 2011 and 2016 most of the FDI inflow was recorded in the real estate sector heavily boosted by the tourism, smart cities, and property development followed by financial and insurance activities.[2]

Several studies have measured the impact of tourism FDI on economic growth, however little attention has been paid on the impact of tourism FDI on sustainable development. There is a need to investigate the dynamic effects of FDI in the tourism sector on sustainable development in Mauritius. Hence, the focus of the paper is mainly to supplement existing literature by investigating this relationship. In this regard, this study uses a rigorous dynamic time series analysis namely a dynamic vector error correction model (VECM), to carry out the proposed investigation. Such a procedure ensures that the dynamic behaviour of the time series under consideration is properly captured, while simultaneously catering for endogeneity and causality issues. Any feedback and indirect effects which might be present will also be detected within the VECM.

The rest of this paper is organised as follows: section "Literature Review" discusses the theoretical and empirical literature; "Methodology" defines the methodological approach used while "Findings" discusses the findings from the study; and finally "Conclusion" presents the conclusions and policy recommendations.

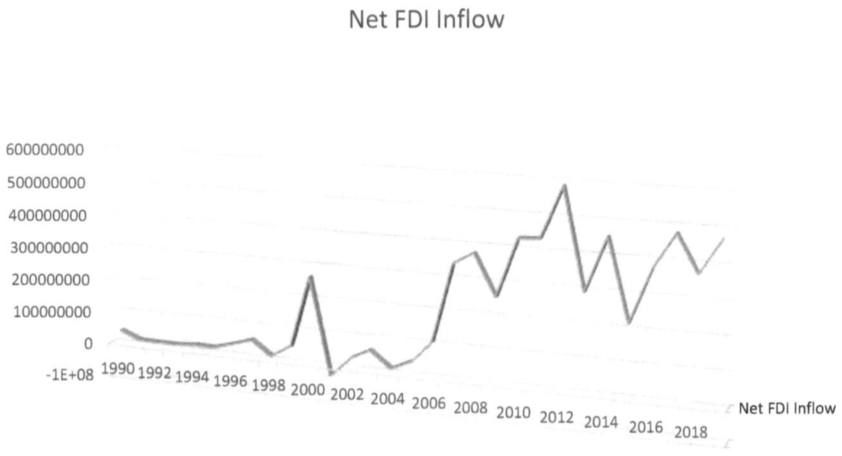

Figure 4.1 Net FDI inflow (US$), Mauritius.

Literature Review

The Concept of Sustainable Development

Sustainable development is mainly achieved when an economy is able to generate development which has an economic, social, and environmental dimension (Rogers et al., 2008). As identified by Grdic and Radic (2012) sustainability requires that the actual generation consume resources in a reasonable manner so that the future generations still have enough resources to maximise their well-being. UNEP (2010) identified sustainability as a link between ecology, economy, and society. Ecology refers to the conservation and management of resources, mainly the non-renewable ones or those crucial for mankind. In terms of economy, reference is made to the level of affluence enjoyed by the community. Finally, society refers to the equal opportunities for all and the respect of human rights. Benefits should be equitably distributed and mainly focus on poverty reduction. Also, different cultures should be respected and all forms of exploitation should be avoided. Hence, sustainable development requires a country to attain the objectives highlighted in Figure 4.2.

Tourism and Sustainable Development

Tourism is recognised as an important sector mainly for developing countries. It is identified as a sector whose benefits are enjoyed by a wider section of the community as compared to other industries in an economy (Tecle and Schroenn, 2006). Many studies have been able to prove the tourism-led growth hypothesis. It refers to the possible links between tourism development and economic growth. This concept is based on the fact that tourism activities have the potential to generate economic

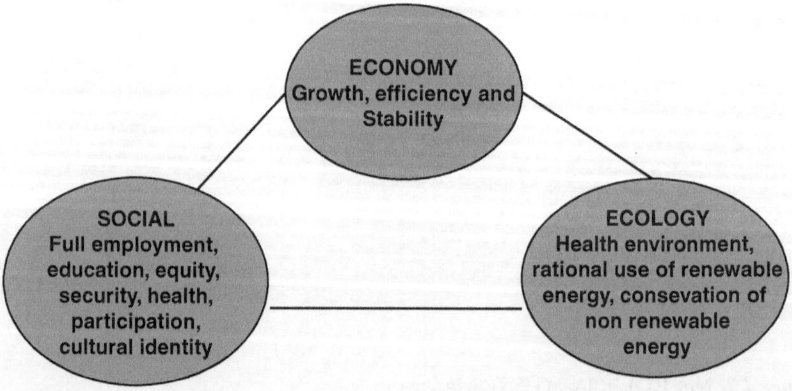

Figure 4.2 Sustainable development.

growth. For instance, Brida and Pulina (2010) have argued that the development of the tourism sector has the possibility to promote economic growth, create employment, and generate tax revenue for the government. The paper was able to prove a bi-directional relationship between tourism and economic growth. Also, expenditure by tourism is considered as an alternate form of exports which provides foreign currency income for an economy which is used to import capital items to produce goods and services. This in turn results in economic growth in the country (Balaguer and Cantavella-Jorda, 2002).

Another strand of the literature shows that tourism flows can generate economic growth through an increase in aggregate demand. Tourists in fact will demand various goods and services ranging from food, transportation, entertainment to accommodation and this will boost production and economic growth. Ashley and Mitchell (2006) and Fauzel (2019), argued that tourism has the capacity to generate social benefits to the host countries. For instance, the concept of pro poor tourism discusses that tourism development leads to poverty alleviation. In addition to that, tourism particularly helps small islands and developing countries having high unemployment, low growth, and other difficulties to competing on an international level (Archer and Fletcher, 1996). The literature thus supports that tourism contributes to the economy and also it brings social benefits to the country. Hence, it benefits consumers and the industry. However, the environmental impact of tourism is debatable. Therefore, while tourism can be very beneficial for sustainable development it can also be very harmful. While tourism brings lots of opportunities to a destination in terms of foreign currencies, employment creation, generating tax revenue for the government, and boosts intercultural indulgence and peace, it can also have negative effects as well. It can, for instance, put pressure on the ecosystem and contribute directly to environmental degradation and cause disturbance to wildlife. The industry can compete to use scarce resources and contribute to local and global pollution. Moreover, this industry represents an unstable source of income and is highly vulnerable. Any external shock might disrupt the flow of international tourism (UNCTAD, 2007). Hence, it is very important to make sure that tourism does not impact adversely on natural resources. Intergenerational equity is very important and thus future generations should not be at a disadvantage because of the current generation.

FDI in Tourism: The Development Dimension

Since tourism represents a lucrative business for island economies mainly, foreign entrepreneurs invest highly in this sector. FDI in the tourism sector is identified as an essential funding avenue. In fact, foreign investment represents an important channel through which capital flows into the host country and also contributes towards uplifting the infrastructural level of the country. More so, there are important knowledge and technological transfers to the destination countries. All these have direct and indirect impacts on the tourism industry (Perić and Radić 2016).

Tourism FDI represents a direct investor who is an entity residing in one country and procures a lasting interest in a tourism related enterprise in another country by engaging in gross fixed capital formation in the tourism sector (UNWTO, 2004). Tourism FDI is noted to occur more frequently in hotels, restaurants, second homes, and passenger transport rental equipment. Also, it occurs occasionally in railway passenger and air passenger transport services whereas in road passenger, water passenger transport services, passenger transport supporting services, travel agencies, and similar cultural services as well as sports and other recreational services, there is less tourism FDI (UNCTAD, 2007). UNWTO (2005) identifies tourism FDI as an engine of growth for tourism companies and a funding source for these firms. However, despite the likely advantages of these firms to the host countries, there may be certain drawbacks as well. There are various factors that influence the flow of these foreign capitals in the country of destination. These include socio-economic environment, economic development level, cultural, historical and geographical distance, hard and soft infrastructure availability, FDI regulations, and privatisation of the industry and company specific factors (Endo, 2006).

Referring to the literature, it is observed that most studies are done on the impact of tourism industry on the economy as a whole or the impact of FDI on the economy. Less studies focusing on the impact of tourism FDI on the economy are found, and even fewer studies on the impact of tourism FDI on sustainable development of the host countries. In order to better investigate the impact of tourism FDI on development, more data is required on several microeconomic indicators or spillovers of tourism FDI such as amount of foreign investment in tourism infrastructure, skills content of employment, technology, and knowledge transfer, among others. Tourism FDI should have an impact at both the macro and micro economic levels in a country. For instance, the UNCTAD (2007) report highlighted the positive and negative effects created by direct and indirect microeconomic linkages between the foreign company and the domestic economy as well as the positive and negative macroeconomic effects on the host countries. The report identified direct effects to be in terms of employment creation in the host country as well as capital expenditure, including accommodation and related infrastructure. Another direct effect is in terms of the foreign company presence in the host country and its brand image. A reputed foreign company can attract tourists to a particular destination. The indirect effects will include mainly the classical consumption multiplier effects whereby the people who have been able to secure employment with the tourism related foreign firms will now spend their income on goods and services from non-tourism related firms thus boosting aggregate demand and hence resulting in an increase in national production and economic growth. The next indirect effect of tourism FDI which was identified by Mirza and Giroud (2004) relates to the value chain multiplier effect mainly through forward and backward linkages with suppliers and distributors. They operate within the host country and internationally as well and thus help the host country to benefit from various advantages in the form of higher level of technology and expertise. More so, there are significant knowledge spillovers as well whereby employees learn from the foreign enterprise and apply these knowledge to local enterprises. Another benefit is in terms of the demonstration effect whereby local hoteliers learn new ways of doing business by merely observing the strategies and choices made by foreign firms. Finally there are also tax contributions made by these firms hence increasing the host country's government revenues.

Apart from the direct and indirect benefits of tourism FDI there are also certain problems associated with such organisations. For instance, there may not be enough benefits which flow to the country as expected. Several leakages are as well identified – precisely, leakage in the national balance of payments in terms of imports of goods or services used in the tourism industry and the repatriation of profits made by foreign firms. Also, tourism FDI can drivc out local firms from the industry. Moreover, these firms may contribute to environmental degradation of the host countries. The extent to which tourism FDI contributes to sustainable development depends on the magnitude of the benefits and complications of this investment on the host countries. Both the positive and negative effects of tourism FDI to the host countries are illustrated in Figure 4.3.

The extent to which tourism FDI contributes to sustainable development is debatable. A crucial feature of sustainable investment in the tourism sector is the economic, social, and environmental viability as highlighted above. The government has an important role to play as an agent of tourism

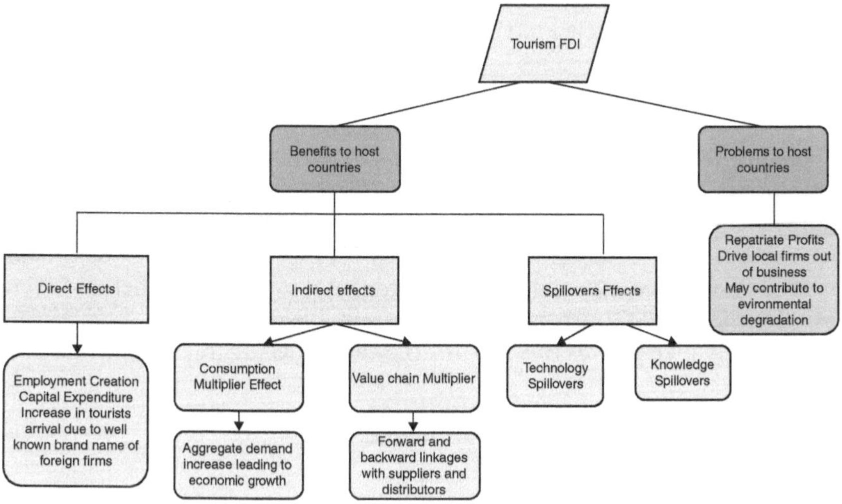

Figure 4.3 Impact of Tourism FDI on host countries.

development. It plays a significant role in the management of the environmental and cultural impacts of tourism. Also, the government is essential in tourism planning and strategy as well as land use planning and upkeep and maintenance of parks, public, and natural attractions (Dwyer et al., 2003). Investment in tourism from all parties does provide much help in developing the tourism sector. Most of the empirical studies show evidence of a positive link between FDI and economic growth. For instance, the paper by Fauzel et al. (2017), using a dynamic VECM, catering for dynamic, endogeneity, and causality issues, addressed the important question of whether FDI in the tourism sector enhances economic growth and productivity in Mauritius using time series data for the period 1988–2013. Their results support the tourism-led growth hypothesis. Other studies investigated the impact of total FDI on economic growth. Zhang (2014) settles that FDI has had large positive effects on China's industrial performance. Other studies found that FDI can have negative impacts on the economy as it leads to the crowding out of domestic investment or even supports unfair competition (Kardos, 2014; Reiter and Steensma, 2010).

Referring to the impact of FDI on environment, there are mixed findings. For instance, the Pollution Haven Hypothesis argues that developed countries invest in developing countries with weak environmental regulations and transfer polluting technologies to these countries (Sarkodie and Strezov, 2019). In their study, Abdouli and Hammami (2020) argue that FDI may have a negative impact on environmental quality by boosting economic growth. However, the Environmental Kuznets curve (EKC) hypothesis (Grossman and Krueger, 1995) stated that economic growth will initially lead to environmental degradation followed by a phase of improvement. Both hypotheses were confirmed by Sarkodie and Strezov (2019); however, it depends on the countries being analysed. The study by Bokpin (2017) revealed that FDI inflows lead to environmental degradation in Africa; nonetheless, the role of governance and institutional policy is crucial in reducing such adverse impacts. On the other hand, Ridzuan et al. (2017) found that FDI inflows do lead to higher economic growth and at the same time, to improved environmental standards.

While the literature is fraught with studies analysing the impact of FDI inflows on economic development and environmental sustainability, less research has been done on the impact of FDI on sustainable development catering for economic, social, and environmental impacts together and even lesser for the case of tourism FDI and sustainable development. Hence, the present study investigates this link for the case of Mauritius over the period 1990–2019.

Methodology

The basic specification of the model used in this study is based on the principles of some earlier studies of growth models carried out by Barro and Lee

(1994), Temple (1999), Durbarry (2004), Seetanah (2011), and Fauzel et al. (2017). However, instead of economic growth, the sustainable development index (SDI) is being used. The econometric model is augmented by including a proxy for tourism development *(TOUR)*, and FDI in tourism *(FDI T)*, FDI non-tourism sector *(FDI NT)*, Human Capital *(HC)*, and Trade Openness *(TO)*.

The model will take the following functional form:

$$\text{Model: SDI} = f\,(\text{TOUR, FDI T, FDI NT, HC, TO}) \tag{4.1}$$

The above model is used to analyse the impact of tourism FDI on sustainable development for the case of Mauritius over the period 1990 to 2019. SDI measures the ecological efficiency of human development while recognising that development must be achieved within planetary boundaries. The SDI was created to update the Human Development Index (HDI) by incorporating the ecological factor in computing the index. Hence, SDI is computed by taking the country's human development score which is made up of life expectancy, education, and income components and dividing it by their ecological overshoot. The ecological component measures the extent to which consumption-based CO_2 emissions and material footprint exceed per-capita shares of planetary boundaries. Countries which have high human development and remain within or near planetary boundaries rise normally to the top.

The SDI formula is as follows:

$$SDI = \frac{Development\ Index}{Ecological\ Impact\ Index}$$

The base formula of the Human Development Index is used with a sufficiency threshold on income. The formula and its components can be described as follows:

$$Development\ Index = \sqrt[8]{\begin{array}{l} Life\ Expectancy\ Index * Education\ Index * \\ Income\ Index \end{array}}$$

The ecological impact index is described as follows:

$$Ecological\ Index = 1 + \frac{e^{AO} - e^{1}}{e^{4} - e^{1}}$$

where *AO* represents the Average Overshoot and is calculated as the material footprint and emissions values divided by their respective per-capita planetary boundary (which varies by year depending on population size) to determine the extent of boundary overshoot (or undershoot). This also standardises the units. If the result of either division is less than 1 (undershoot) it is rendered as 1. Then the results are averaged using the geometric mean. This method ensures that a country cannot compensate for overshooting one boundary by undershooting the other. Overshoot of either boundary will yield average overshoot of greater than 1.[3]

The variable of interest for this paper is FDI in the tourism sector. The extent of foreign presence in the tourism sector is measured by *FDI T* and the proxy used is FDI in the tourism sector as a percentage of real GDP. Data for FDI flowing in the tourism sector for Mauritius is extracted from the Balance of Payments reports as provided by the Bank of Mauritius. There are several benefits that a country derives from foreign companies. For instance, it can boost the country's economic growth as foreign tourism companies act as a promoter for the injection of fresh capital in the host country and also help in attracting foreign tour operators and tourists (Yunis, 2008). Other scholars such as Dwyer et al. (2003) argue that foreign investment and know-how are considered to be of paramount importance in creating and upgrading tourism-related infrastructure and also foreign investment can give rise to more investment in tourism in total.

In order to investigate the impact of non-tourism FDI (*FDI NT*) on sustainable development, FDI in the non-tourism sector is included in the model. Furthermore, tourism development as proxied by the number of tourist arrivals is incorporated in the model. For instance, considering the tourism-led growth hypothesis, the papers by Durbarry (2004), Louca (2006), Noriko and Mototsugu (2007), and Gani (1998) support the positive relationship between international tourism and economic growth for small island economies using co-integration and causality tests. Hence, by including this variable it will help to detect whether tourism development also contributes to sustainable development. The data was extracted from the Statistics Mauritius database.

Furthermore, Human capital (*HC*) as a measure of education level is added in the model. A population which is more educated will be better able to contribute towards a sustainable world. A good education system is of paramount importance to promote sustainable development. Education for sustainable development contributes towards the development of the knowledge, skills, understanding, values, and actions required to create a sustainable world, which ensures environmental protection and conservation, promotes social equity, and encourages economic sustainability (UN, 2002). Hence, secondary enrolment ratio has been included in the study and the data has been extracted from Statistics Mauritius. In addition to that, scholars have been arguing that trade policies remain a focal point for both direct and indirect relations with sustainable development. On one hand side, trade can generate economic growth, create employment opportunities as well as ensure efficient resource allocation required for economic development. Trade has been recognised as a tool to promote sustainable development (Balassa, 1986). As argued by GATT Director-General Arthur Dunkel and reported by Weiss (1992), trade is not an end to environmentally sustainable economic development, but rather a powerful enabler or engine. Hence, to investigate the impact of trade on sustainability in Mauritius, the variable trade openness (*TO*) is included in the model. Trade openness is measured as the sum of imports and exports to GDP.

In order to ease interpretation of the results, the natural logarithm of the variables has been employed. The model is therefore rewritten as follows:

$$lnSDI_t =_1 lnFDIT_t +_2 lnFDINT_t + {}_3 lnTOU_t + \beta_4 lnHC_t + \beta_5 TO_t + \mu_t \quad (4.2)$$

where t denotes the time dimension and *ln* represents the natural logarithm and the variables are as previously defined.

Estimation Issues

Applying regression on time series data may generate spurious results (Granger and Newbold, 1974; Philips, 1986) due to the possibility of non-stationarity data. Hence, checking the stationarity of data is a prerequisite for applying co-integration test. As a result the Phillips-Perron test (Phillips and Perron 1988) was applied. Once the variables are stationary of the same order, the second step is to check for co-integration test or long run co-integration relationship amongst the variables. The Johansen Co-integrating Test (Johansen 1988; Johansen and Juselius, 1990), which uses maximum likelihood testing process, is applied, to investigate the number of Co-integration vectors in the Vector Auto Regressive (VAR) setting.

Actually, the static single equation often fails to take into account the presence of dynamic feedback among relevant variables. Accordingly, a VAR approach to study the relationship between FDI in tourism and sustainable development is chosen. Such an approach does not impose *a priori* restriction on the dynamic relations among the different variables. It resembles simultaneous equation modelling, whereby several endogenous variables are considered together.

The common form of VAR is given as follows:

$$Z_t = \lambda + \gamma_t Z_{t-1} + \cdots + \gamma_k Z_{t-k} + \varepsilon_t \quad (4.3)$$

Where Z is an $(n \times 1)$ vector of k variables having integrated of order 1 that is $I(1)$, λ is a $(n \times 1)$ vector of intercepts, $\gamma_t, \ldots \gamma_{t-k}$, are parameters and ε_t is a normally distributed residual term. The common VAR based model in equation (4.3) may also take the form of the VECM as follows:

$$\Delta Z_t = \lambda + \Gamma \Delta Z_{t-1} + \prod Z_{t-1} + \varepsilon_t \quad (4.4)$$

$$Z_t = [lnSDI_t, lnFDIT_t, lnFDINT_t, lnTOU_t, lnHC_t, lnTO_t]$$

Where Z is an $(n \times 1)$ vector of k variables having integrated of order 1 that is $I(1)$, λ is a $(n \times 1)$ vector of intercepts, ε_t is an $(n \times 1)$ vector of residuals. Further, Δ is the difference operator and Γ and $\prod$ are coefficient matrices.

Π is also known as the impact matrix as it explains the long run equilibrium relationship of the variables; while Γ explains the short run effect. The VECM linking short term and long term causality between Tourism FDI and sustainable development is set as follows;

The coefficient of the error correction term (ECT_{t-1}) indicates whether there exists a short run relationship among the time series variables. The sign and value of the coefficient provides information about the speed of convergence or divergence of the variables from their long run co-integrating equilibrium. A negative and significant coefficient of ECT_{t-1} is favourable for the stability of long run equilibrium.

Findings

The Phillips-Perron (PP) (1988) unit-roots tests were employed and the results of the stationarity tests suggest that all our variables are integrated of order 1 and stationary in first difference. The Johansen Maximum Likelihood approach is subsequently used to test the presence of co-integration in a VECM. Trace statistics λ_{trace} and maximal eigenvalue are used to check the number of co-integrating vectors. These statistics test the null hypothesis of no co-integrating equations against the alternative of co-integration. The trace statistics either reject the null hypothesis of no-cointegration among the variables or do not reject the null hypothesis that there is a co-integration relation between the variables. The results show the presence of co-integrating vectors and we thus conclude that a long run relationship exists between the dependent variable and the control variables.

Empirical Results

Vector Error Correction Model, VECM

Since there is the presence of co-integration, and hence a long run equilibrium relationship among the variables has been established, the next step is to specify and estimate a VECM including the error correction term to investigate the dynamic nature of the model. The VECM specification forces the long run behaviour of the endogenous variables to converge to their co-integrated relationships, which accommodates short run dynamics. In this study, the VECM is estimated using an optimum lag length of 1. The long run coefficients are reported in Table 4.1.

The long run equation provides very stimulating results. Analysing Table 4.1, the impact of FDI in tourism on SDI in Mauritius, it is observed that it has indeed contributed to sustainable development in the economy as supported by the positive and significant coefficient of the variable. In fact a 10% increase in FDI in tourism raises sustainable development by 0.24%. This result is in line with Aust et al. (2020). FDI in the tourism sector in Mauritius started to increase from the year 1995. AHRIM (Association of

Table 4.1 Long run coefficients

	Co-integration equation 1	*t statistics*
lnSDIt	1.00000	
lnFDITt	0.024235***	3.55840
lnFDINTt	−0.173178***	−9.74548
lnTOUt	0.626688***	3.89444
lnHCt	0.542908***	1.83892
lnTOt	1.007327***	7.22126
Constant	14.96997	

*** represents significance at 1%.

Hoteliers and Restaurants in Mauritius, 2017: https://www.ahrim.mu/publications/annual-report/), pointed out that entrepreneurs in the tourism sector have realised that competitiveness and sustainability of the tourism industry work together as the quality of tourist destinations is highly influenced by their natural and cultural environment and their integration into the local community.

Interestingly it is noted that tourism development has contributed to sustainable development in the country. More precisely, a 1% increase in tourism development has led to 0.363% increase in sustainable development. Importantly Mauritius has achieved the "GSTC-Recognized Standard" status as announced by the Global Sustainable Tourism Council (GSTC). The Mauritian Standard on Sustainable Tourism (MS 165:2019) was established to guide the sustainable development of the tourism industry in Mauritius. It targets tourism business and non-tourism related activity in order to promote its sustainability performance. The aim of the Sustainability Tourism Standard is to address requirements of the environmental impacts and its effect on land, air, water, and other organisms and ecosystem of the island.[4]

The results further support the view that education has contributed to sustainable development. As highlighted by Bourn et al. (2017), secondary education ensures that learners have free access, equitable, and quality education promoting the Sustainable Development Goal (SDG) 4. In fact, education does promote the development of knowledge, skills, understanding, values, and actions which are all necessary for a sustainable world, hence ensuring the protection of the environment above all. Moreover, Mauritius strives to have an education system that promotes the holistic development of all citizens. In addition to that, the aim is to achieve 100% MID (Maurice Ile Durable (Mauritius Sustainable Island)) literacy by 2020 and be an internationally recognised knowledge hub for sustainable development in the region by 2020.[5]

Finally, focusing on the trade openness result, it can be found that even this variable is positive and significant as per the current study. Similar results have been reported by Antweiler et al. (2001) and Frankel and Rose (2005) who found that trade is not necessarily bad for the environment.

This result is, however, against environmentalists and dependency theorists who claim that trade openness is in contradiction to sustainable development. It is also in contradiction to the study of Sheikh et al. (2020), who used an autoregressive lag (ARDL) model and investigated the impact of trade openness on sustainable development for India. Their results shows that trade openness is distorting and detrimental to future generations.

However, referring to the non-tourism FDI results, it is seen that it deteriorates sustainable development in the country. In fact, Mauritius is actively working towards implementing the SDGs and recognises them to be a key to development. In a mapping exercise that the country did in 2016, 222 indicators were identified and integrated in the national policy. However, the country is still working towards a permanent national structure to monitor continual progress for SDGs.[6]

Hence, in line with the results, it can be confirmed that tourism FDI has contributed to sustainable development and the study confirms the presence of sustainable tourism development in the country. More so, education as well as trade openness has contributed to sustainable development. However, non-tourism FDI does not contribute towards sustainable development.

Conclusion

Based on the VECM framework adopted in this study, the main objective is to investigate the relationship that might exist between tourism FDI and sustainable development in Mauritius over the period 1990–2019. Referring to the results, it is observed that tourism FDI does result in sustainable development in Mauritius in the long run. Actually, the government of Mauritius has adopted liberalising policies to boost FDI in the tourism sector. Significant amount of foreign investment has been registered in key tourism activities such as restaurants, yachts, and travel agencies among others. These investments have contributed towards the development of the country as there have been important spillover effects in terms of capital flows, construction of world-class hotels and villas, as well as creation of employment. This argument can partially explain the results obtained for tourism FDI and sustainable development.

Also, the results show that the island has sustainable tourism development and both education and trade openness have contributed to the sustainable development of the country. However, non-tourism FDI is noted to be detrimental to sustainable development. Overall, this study provides new evidence in the field of tourism FDI and sustainable development for the case of Mauritius using a recent co-integration approach in a dynamic framework. However to further boost sustainable development in Mauritius, there are various actions that can still be taken. For instance, it is noted that there is bad governance of natural resources in the tourism sector. Large scale developments are being approved for smart cities and IRS (Integrated Resort Scheme) on areas rich in flora and fauna. Also, there is a significant amount of pollution generated by the tourism sector through massive water sports investment and sea activities invading animals in their natural habitats. It is important therefore to come up with eco-conscious policies to counteract the damages to the island. Water activities and other sea sports need to be done in a more conscious and eco-friendly manner. The 2016 Conscious Travel Sustainability Assessment greatly helps to attain these objectives but constant effort is needed to further improve sustainable investment in the tourism sector.

Notes

1 https://wttc.org/Research/Economic-Impact.
2 https://unctad.org/system/files/official-document/diaepcb2017d9_en.pdf.
3 https://www.sustainabledevelopmentindex.org/.
4 https://www.gstcouncil.org/the-mauritian-standard-on-sustainable-tourism-is-gstc-recognized/.
5 https://sustainabledevelopment.un.org/content/documents/1109215 Mauritius%20National%20Report.pdf.
6 https://sustainabledevelopment.un.org/memberstates/mauritius.

References

Abdouli, M. and Hammami, S., 2020. Economic growth, environment, FDI inflows, and financial development in Middle East countries: Fresh evidence from simultaneous equation models. *Journal of the Knowledge Economy*, *11*(2), pp. 479–511.

Archer, B. and Fletcher, J., 1996. The economic impact of tourism in the Seychelles. *Annals of Tourism Research*, *23*(1), pp. 32–47.

Ashley, C. and J. Mitchell, 2006. Can Tourism Help Working Papers, Fondazione Eni Enrico Mattei. Reduce Poverty in Africa?, Overseas Development 29.Antweiler, W., Copeland, B.R. and Taylor, M.S., 2001. Is free trade good for the environment? *American Economic Review*, *91*(4), pp. 877–908.

Aust, V., Morais, A.I. and Pinto, I., 2020. How does foreign direct investment contribute to sustainable development goals? Evidence from African countries. *Journal of Cleaner Production*, *245*, p. 118823.

Balaguer, J. and Cantavella-Jorda, M., 2002. Tourism as a long-run economic growth factor: The Spanish case. *Applied Economics*, *34*(7), pp. 877–884.

Balassa, B., 1986. Policy responses to exogenous shocks in developing countries. *American Economic Review*, *5*, pp. 75–78.

Barro, R.J. and Lee, J.W., 1994. Sources of economic growth. In *Carnegie-Rochester Conference Series on Public Policy* (Vol. 40, pp. 1–46). North-Holland.

Becker, C. and George, B.P., 2011. Rapid rail transit and tourism development in the United States. *Tourism Geographies*, *13*(3), pp. 381–397.

Bokpin, G.A., 2017. Foreign direct investment and environmental sustainability in Africa: The role of institutions and governance. *Research in International Business and Finance*, *39*, pp. 239–247.

Bourn, D., Hunt, F. and Bamber, P., 2017. *A Review of Education for Sustainable Development and Global Citizenship Education in Teacher Education*. Paris: UNESCO.

Brida, J. and Pulina, M., 2010. *A Literature Review on the Tourism-Led-Growth Hypothesis.* Italy: Centre for North South Economic Research, University of Cagliari and Sassari, Sardinia.

Briedenhann, J. and Wickens, E., 2004. Tourism routes as a tool for the economic development of rural areas—vibrant hope or impossible dream? *Tourism Management, 25*(1), pp. 71–79.

Durbarry, R., 2004. Tourism and economic growth: The case of Mauritius. *Tourism Economics, 10*(4), pp. 389–401.

Dwyer, L., Forsyth, P., Spurr, R., & Vanho, T., 2003. Tourism's contribution to a state economy: A multi-regional general equilibrium analysis. *Tourism Economics, 9*(4), pp. 431–448.

Endo, K., 2006. Foreign direct investment in tourism – Flows and volumes. *Tourism Management, 27*, pp. 600–614.

Fauzel, S., 2016. Tourism and employment spillovers in a small island developing state: A dynamic investigation. *Theoretical Economics Letters, 6*(02), 138.

Fauzel, S., 2019. Tourism and income inequality in Mauritius: An empirical investigation. In edited by Dogan Gursoy and Robin Nunkoo, *The Routledge Handbook of Tourism Impacts* (pp. 109–120). London: Routledge.

Fauzel, S., Seetanah, B. and Sannasee, R.V., 2015. Productivity spillovers of FDI in the manufacturing sector of Mauritius. Evidence from a dynamic framework. *The Journal of Developing Areas, 49*(2), pp. 295–316.

Fauzel, S., Seetanah, B. and Sannassee, R.V., 2017. Analysing the impact of tourism foreign direct investment on economic growth: Evidence from a small island developing state. *Tourism Economics, 23*(5), pp. 1042–1055.

Frankel, J.A. and Rose, A.K., 2005. Is trade good or bad for the environment? Sorting out the causality. *Review of Economics and Statistics, 87*(1), pp. 85–91.

Gani, A., 1998. Macroeconomic determinants of growth in the South Pacific island economies. *Applied Economics Letters, 5*, pp. 747–749.

Granger, C. and Newbold, P., 1974. Spurious regressions in econometrics. *Journal of Econometrics, 2*(2), pp. 111–120.

Grdic, Z.S. and Radic, M.N., 2012. Foreign direct investment and development of sustainable tourism: Case of Republic of Croatia. In *Faculty of Tourism and Hospitality Management in Opatija. Biennial International Congress. Tourism & Hospitality Industry* (p. 447). University of Rijeka, Faculty of Tourism & Hospitality Management.

Grossman, G.M. and Krueger, A.B., 1995. Economic growth and the environment. *The Quarterly Journal of Economics, 110*(2), pp. 353–377.

Hindley, B. and Smith, A., 1984. Comparative advantage and trade in services. *World Economy, 7*(4), pp. 369–390.

Johansen, S., 1988. Statistical analysis of co-integrating vectors. *Dynamics and Control, 12*(2–3), pp. 231–254.

Johansen, S. and Juselius, K., 1990. Maximum likelihood estimation and inference on co-integration-with applications to the demand for money. *Oxford Bulletin of Economics and Statistics, 52*(2), pp. 169–210.

Kardos, M., 2014. The relevance of foreign direct investment for sustainable development. Empirical evidence from European Union. *Procedia Economics and Finance, 15*, pp. 1349–1354.

Lew, A.A., 2011. Tourism's role in the global economy. *Tourism Geographies, 13*(1), pp. 148–151.

Liasidou, S., 2013. Decision-making for tourism destinations: Airline strategy influences. *Tourism Geographies*, *15*(3), pp. 511–528.

Louca, C., 2006. Income and expenditure in the tourism industry: Time series evidence from Cyprus. *Tourism Economics*, *12*(4), pp. 603–617.

Mihalič, T., 2014. "Chapter 3: Tourism and Economic Development Issues". In edited by Richard Sharpley and David J. Telfer, *Tourism and Development (2nd Ed): Concepts and Issues* (pp. 77–117). Bristol, Blue Ridge Summit: Channel View Publications, 2014. https://doi.org/10.21832/9781845414740-005

Mirza, H. and Giroud, A., 2004. Regionalization, foreign direct investment and poverty reduction: Lessons from Vietnam in ASEAN. *Journal of the Asia Pacific Economy*, *9*(2), pp. 223–248.

Narula, K., 2012. 'Sustainable Investing' via the FDI route for sustainable development. *Procedia-Social and Behavioral Sciences*, *37*, pp. 15–30.

Noriko, I. and Mototsugu, F., 2007. Impacts of tourism and fiscal expenditure to remote islands: The case of the Amami islands in Japan. *Applied Economics Letter*, *14*, pp. 661–666.

Perić, J. and Radić, M.N., 2016. FDI-led tourism growth hypothesis: Empirical evidence from Croatian tourism. *European Journal of Tourism, Hospitality and Recreation*, *7*(3), pp. 168–175.

Philips, P.C.B., 1986. Understanding spurious regression in econometrics. *Journal of Econometrics*, *33*(3), pp. 311–340.

Phillips, P.C. and Perron, P., 1988. Testing for a unit root in time series regression. *Biometrika*, *75*(2), pp. 335–346.

Rogers, P.P. and Jalal, K.F.C., 2008. *An introduction to sustainable development* (No. 338.927 R6329i Ej. 1 022615). EARTHSCAN.

Reiter, S.L. and Steensma, H.K., 2010. Human development and foreign direct investment in developing countries: The influence of FDI policy and corruption. *World Development*, *38*(12), pp. 1678–1691.

Ridzuan, A.R., Ismail, N.A. and Che Hamat, A.F., 2017. Does foreign direct investment successfully lead to sustainable development in Singapore? *Economies*, *5*(3), p. 29.

Sarkodie, S.A. and Strezov, V., 2019. Effect of foreign direct investments, economic development and energy consumption on greenhouse gas emissions in developing countries. *Science of the Total Environment*, *646*, pp. 862–871.

Seetanah, B., 2011. Assessing the dynamic economic impact of tourism for island economies. *Annals of Tourism Research*, *38*(1), pp. 291–308.

Sheikh, M.A., Malik, M.A. and Masood, R.Z., 2020. Assessing the effects of trade openness on sustainable development: Evidence from India. *Asian Journal of Sustainability and Social Responsibility*, *5*(1), pp. 1–15.

Tecle, Y.H. and Schroenn, J.L., 2006. The contribution of HRD to tourism-led development in an African context: Economics. *South African Journal of Economic and Management Sciences*, *9*(4), pp. 444–457.

Temple, J., 1999. The new growth evidence. *Journal of Economic Literature*, *37*(1), pp. 112–156.

UNCTAD, 2007. *World Investment Report 2007: Transnational Corporations, Extractive Industries and Development*. New York: United Nations.

UNEP, 2010. Study on mercury sources and emissions, and analysis of cost and effectiveness of control measures "UNEP Paragraph 29 study". United Nations

Environmental Programme, Chemicals Branch, DTIE, Geneva, Switzerland, November 2010.

United Nations, 2002. *Report of the World Summit on Sustainable Development*, Johannesburg, available: http://daccessdds.un.org/doc/UNDOC/GEN/N02/636/93/PDF/N0263693.pdf?OpenElement.

UNWTO, 2001. *Tourism 2020 Vision*. Madrid: UNWTO.

UNWTO, 2004. *Tourism Market Trends*. Madrid: UNWTO.

Weiss, E.B., 1992. Environment and trade as partners in sustainable development: A commentary. *American Journal of International Law, 86*:728–735

WTO Commission for Africa. Meeting, 2005. *Tourism Market Trends, Africa*. World Tourism Organization.

Yunis, E., 2008. Attracting FDI with good CSR practices in the tourism sector. In *Responsible Enterprise, Foreign Direct Investment and Investment Promotion* (p. 99). International Institute for Environment and Development.

Zhang, K.H., 2014. How does foreign direct investment affect industrial competitiveness? Evidence from China. *China Economic Review, 30*, 530–539.

5 Institutional Influences on National Tourism-Related Foreign Direct Investment (TFDI) Policies and Programmes

Kyrgyzstan, Central Asia

Nicola J. Palmer

Introduction

This chapter critically examines tourism-related foreign direct investment (TFDI) in the context of Kyrgyzstan, Central Asia through an evolutionary lens. The chapter focuses on the case of the Pinara-Bishkek Hotel, one of the first foreign hotels to be established in the republic following the dissolution of the USSR in 1991. It is intended that this chapter will provide insights into the way in which a TFDI develops and evolves over time acknowledging the dynamic nature and complexity of this type of investment. Secondary data has been collected from a range of publicly available online sources and qualitative document analysis is applied to enable process tracing. The evolutionary path of a TFDI in Kyrgyzstan is explored and consideration is given to how this might inform wider understanding of sustainable tourism development, particularly in a post-Soviet Central Asian context.

Within the context of national development strategies, foreign direct investment (FDI) has received considerable attention from policymakers and academics. Policies and programmes relating to FDI require decision-making and the involvement and cooperation of a range of formal (and informal) institutions. This is particularly noted in respect to TFDI – see for example, Dwyer's (2014) consideration of power and influence between TFDI "players". His analysis draws attention to the directional tendencies of globalisation, economic interdependence and the multidimensional process involved in increased global connectivity and global consciousness that combines society, economy, and culture. The relevance to sustainability – viewed traditionally as "an economic phenomenon associated with the development of the global market" (Parliamentary Assembly, 2003: 1) – is clear.

With respect to globalisation, the potential of FDI to facilitate host country integration into international tourism networks, increasing tourist flows and income from tourism-related activities has been noted by Endo (2006). Perić and Radi

DOI: 10.4324/9781003155492-7

(2011: 263) argue:

> Developing countries, if they wish to promote sustainable development in the context of tourism, should be extremely cautious in attracting FDI in this sector. It is necessary to attract sustainable FDI.

Research on FDI and TFDI has often overlooked or neglected the random or fortuitous nature of political, economic, and social outcomes. It has also failed to appreciate the challenges faced by transitional economies with limited development options, fragile geopolitical ties (Elhawary et al., 2010), and the influence of the past.

According to the Organisation for Economic Co-operation and Development (OECD, undated: 7):

> Foreign direct investment (FDI) is a category of investment that reflects the objective of establishing a lasting interest by a resident enterprise in one economy (direct investor) in an enterprise (direct investment enterprise) that is resident in an economy other than that of the direct investor. The lasting interest implies the existence of a long-term relationship between the direct investor and the direct investment enterprise and a significant degree of influence on the management of the enterprise.

It is interesting to explore the evolution of TFDI in a transitional economy and the potential implications for sustainable tourism development.

Literature Review

Much of the research that exists on TFDI focuses on quantitative measurement studies. It adopts a structured approach, often directed along the guidelines of financial policy "headlines" and is focused on investment impacts. Drope et al.'s (2014) exploration of the political economy of FDI, for example, identifies three key impact themes: tax reforms; trade agreements; industry regulation. However, there is merit in returning to Dwyer's (2014) ideas and exploring the socio-cultural and power aspects of TFDI beyond the obvious economic relevance (Long, 1997; Lucke and Eichler, 2016; Donaldson and Forssman, 2020). There is also value in considering the institutional and historical context of TFDI and the boundaries of time and space that influence the trajectory of development when examining the sustainability implications of TFDI at a national level.

The Institutional and Cultural Context of TFDI

Institutional and cultural characteristics have been recognised to be influential factors in quantitative analyses of TFDI. There have been some interesting propositions made about the relationships between investor country and host country culture with respect to "cultural proximity". Lucke and Eisler (2016: 935), for example, have claimed that:

> institutional and cultural distance is important and that FDI has a predominantly regional aspect. FDI to developing countries is positively affected by better institutions in the host country, while foreign investors prefer to invest in developed countries that are more corrupt and politically unstable compared to home. The results indicate that foreign investors prefer to invest in countries with less diverse societies than their own.

The work of Lucke and Eisler (2016) draws attention to the agency of foreign investors. Less attention has been paid to both host government and investor actors. Moreover, quantitative analyses seeking to produce generalisations have dominated academic discourse on TFDI. Qualitative studies offer an opportunity to gain a richer picture of TFDI relations. Those studies have, however, to date, tended to concentrate on societal impacts, often focusing on resident perceptions and experiences. They have also tended to provide "snapshot" pictures rather than exploring lasting interests.

There is scope to learn from policy and planning analyses of tourism relationships and changes over time. In their analysis of tourism institutions, Mellon and Bramwell (2018: 42) argue that "a fuller understanding of tourism processes should include analysis of historical influences, legacies and the sequencing of change." They bring together the theories of Historical Institutionalism (linked to Institutional Theory and Evolutionary Economic Geography) and Cultural Political Economy to understand institutional change via a process tracing methodology. Evolution may also be seen to hold relevance to achieving deeper insights into TFDI within the context of national-level development.

One of the most common approaches to examining evolution in the context of tourism studies and destination development is Evolutionary Economic Geography (EEG) (Brouder and Eriksson, 2013). Brouder and Eriksson (2016:385) argue that, "Tourism scholars who engage with EEG will discover fertile ground for inquiry by addressing the roles of enterprise, networks, and the state in shaping regional evolution." Boschma and Martin (2010) outline three major theoretical frameworks for EEG: Generalized Darwinism – drawing on concepts from modern evolutionary biology; Complexity Theory – concerning aspects of "far-from-equilibrium" adaptive systems; Path Dependence Theory – based on the role of contingency and self-reinforcing (autocatalytic) dynamics (Ma, 2013). EEG draws attention to "bounded rationality" and "routines" (Simon, 1955) in the exploration of the behaviours of organisations. This approach may offer a means of gaining insight into both host government investment economy and foreign investor behaviours and actions over a sustained time period. As Brouder and Eriksson (2016: 384) acknowledge, "understanding how the economy evolves enhances our ability to question the nature of that evolution with, for example, growth-oriented models often challenged as unsustainable in tourism (Butler, 1999)."

Pike et al. (2016: 127) purport that Geographical Political Economy (GPE) is a useful approach to political-economic and evolutionary thinking and "contingency and particularity generated by agency and context can be handled through the techniques of 'following the path' and 'deep con textualisation." They advocate consideration of how and why "specific paths unfold in particular ways over time and space" (Pike et al., 2016: 131) and this requires consideration of cultural aspects of institutional context recognising the existence of beliefs, values, and practices surrounding agency in decision-making.

Development Paths and Challenges Faced by Transitional Economies

EEG, Historical Institutionalism, and GPE are concerned with development paths. Brouder and Eriksson (2013: 378) note that:

> path dependence is the term used to reflect the inertial trajectory of a region as a result of long-term processes, a state which will only be altered by either major intervention, some external shock, or the embedding of new seeds of structural change in the hope that they will germinate and grow in the long-term.
>
> (Neffke and Svensson Henning, 2010)

Debates on the ability of places to change path dependence have raised question marks over the influence of historical processes on the development of alternative paths (Henning et al., 2013) and the constraints that affect agency and the ability to break with path trajectories. As Brouder and Eriksson (2016: 379) identify, there is a need for wider understanding, not least in terms of *how* tourism development is "introduced to regions which were previously reliant on different sectors, e.g., resource-based economies in peripheral regions" as is the case with Kyrgyzstan.

The concept of "inclusive sustainable development" and, arguably, sustainable tourism development is dependent on a global policy environment that is "conducive to cross-border investment" (UNCTAD, 2019: 6). This environment is challenged by differing levels of development between countries and a need for external assistance has been acknowledged (Edwards et al., 1999; Mitlin et al., 2007). Slocum and Backman (2011: 281) have argued that, "good governance is a prerequisite in achieving sustainability objectives." In relation to this Nørgaard (2000) has identified the significance of external interventions in the democratic reform of post-communist countries, noting "the virtuous and vicious circles of emerging institutions."

The role that conflict and contestation (Mahoney and Thelan, 2009; Conran and Thelen, 2016) play in macro-level TFDI agreements at national level has been acknowledged. There are certainly specific challenges to globalisation faced by developing and transitional economies and value can be attributed to the interventions of external actors. Werner's (2003) exploration of tourism development in Central Asia highlights the importance of tourism mediators in the face of a lack of well-developed tourist infrastructure and acknowledges their role in "cultivating a positive image of Central Asia as a new tourist destination, developing tourist accommodations, and lobbying government institutions to support and regulate tourism" (Werner, 2003: 141).

In terms of "developing tourist accommodations", comprehensive research by the United Nations Conference on Trade and Development (UNCTAD, 2004) highlights that hotels and restaurants dominate TFDI activities in both developed and developing countries. A heavy focus on the accommodation sector in TFDI has been noted (Endo, 2006; Barrowclough, 2007; UNCTAD, 2010) and the employment of tourism as a political tool (Henderson, 2003, 2011) has also been noted in this context (Khoshnevis et al., 2017).

One reason for dominance of the accommodation sector in TFDI in developing and transition countries relates to FDI development strategy advice from international organisations (Perić and Radić, 2011) and international production factors relating to ownership advantages, location advantages, and internalisation advantages (Dunning and Mcqueen, 1981).

TFDI Decisions and Investment Opportunities

The issue of who is in the host economy space in terms of influencing development decisions, linked to investment opportunities, certainly warrants attention. Development decisions are not solely determined by national governments.

As Endo (2006) and Perić and Radić (2011) acknowledge, determinants of TFDI do not differ from those of other industries and include:

> cultural/historical/geographical distance, political and/or economic risks, level of economic development, socio-economic environments, privatization of the industry, liberalization of FDI regime, taxation, investment incentives, availability and quality of hard and soft infrastructures and corporate strategies or company-specific factors.
>
> (Endo, 2006: 601)

Political regimes have been noted to be influential in the attraction of TFDI and the specific nature of TFDI relationships. From a host country perspective, Khoshnevis et al. (2017) acknowledge the presence of hope and fear surrounding TFDI and its offerings versus its potential economic, cultural, community, and environmental impacts.

Much research relating to FDI and TFDI is grounded in political economy. Revisiting Britton's (1982) seminal work on the political economy of tourism is useful in relation to the tourism aspect of TFDI. Conceptualising international tourism as "a product of metropolitan capitalist enterprise" (Britton, 1982: 331) it is important to reflect on the low level of control that a nation has in a global system. The destination area as an attractor has sustained attention in the tourism literature (Plog, 1974; Butler, 1980; Hu and Ritchie, 1993; Hu and Wall, 2005; Crouch, 2011; Andrades and Dimanche, 2017; Nazmfar et al., 2019). Inevitably, TFDI is driven by investors' perceived potential for tourism revenues to be generated within the destination area.

Conversion of tourism promotion activities to investment and visitation can be difficult in lesser-known tourist destinations. Fauzel et al. (2017: 1044) acknowledge the potential contribution that one foreign investor in tourism can make to the overall TFDI in a country: "foreign tourism companies also often act as catalysts for the injection of fresh capital in the host country and help in attracting foreign tour operators and tourists." Baidoo (2018) has also emphasised the potential for foreign tourist arrivals to increase FDI inflows due to a discovery of investment opportunities during visitation. The destination experiences of initial inward investors and inbound visitors, particularly in the context of Central Asia (Lee et al., 2012) characterised by lesser-known tourist destinations, appear to offer reputation-building benefits to assist wider national economic growth. Looking at perceptions of Turkish investors on the tourism market in Central Asia, including Kyrgyzstan, Kantarci (2007: 828) noted the importance of developing "country-specific investment strategies and incentives to attract more foreign investment in tourism industries." He suggested that "the officials of …[Central Asian] countries could use currently active companies in their respective countries to encourage them to team up with other companies to further attract more investment capital" (Kantarci, 2007: 827). This may be an important strategy given Copeland's (1991) caution that excessive trust and a high risk to the host community may result from too much FDI.

Cultural distance and risk reduction strategies have been acknowledged in relation to international tourism visitor flows (Litvin et al., 2004; Ahn and McKercher, 2015) as well as TFDI inward investment decisions (Sanford and Dong, 2000; Quer et al., 2007; Deng et al., 2019). These link to cognitive and behavioural issues in finance (Dincer et al., 2016) and the importance of considering both structure and agency when considering the impacts of tourism investment flows on tourism development (Bianchi, 2003; Meyer, 2013).

The Institutional-Historical Context of Kyrgyzstan

Located in Central Asia, Kyrgyzstan is a land-locked, lower middle-income transition economy (Brück and Esenaliev, 2018).

The country gained independence from the USSR in 1991 alongside neighbouring Commonwealth of Independent States (CIS) countries,

Kazakhstan, Tajikistan, and Uzbekistan. Geographically, the country is distinctive from other parts of Central Asia due to its size, topography, and approach to democratic reform. Ranked as one of the poorest Central Asian countries, the country has an area of 198,500 km^2 (76,641 mi^2) and with an estimated 6.5 million population in 2019 (World Bank, 2020b), it is one of the world's most sparsely populated countries. The traditional nomadic culture of the Kyrgyz has shifted towards settlement and it has been reported that 35.6% of the population now reside in urban areas, namely Bishkek in the north and Osh in the south (Worldometer, 2021).

Over 90% of the land area in Kyrgyzstan is mountainous and the scenery has repeatedly prompted the accolade "the Switzerland of Central Asia". However, lack of ease of accessibility, under-developed infrastructure together with ongoing political and economic instability, ethnic tensions, threat of religious extremism, and volatile borders restrict international tourism growth (Palmer, 2014).

The country's capital city Bishkek (formerly named *Pishpek* and *Frunze*) was formally established as the country's political and administrative centre in 1936 during the final stages of the national delimitation (*razmezhevanie*) in the Soviet Union. It holds economic significance and is second to Lake Issyk-Kul (*Ysyk-Köl*) in terms of tourism importance. Bishkek, located ten miles from the border with Kazakhstan and 25 km (16 miles) south southwest of Manas International Airport is home to the country's only international hotels and two Western-style supermarkets. The city may be identified as the locus of international trade in Kyrgyzstan.

Kyrgyzstan's Trading Context

Allayarov et al. (2018: 95) have claimed:

> Kyrgyzstan, being a less developed economy, even by Central Asia standards, can only achieve its goals of reducing poverty and becoming more developed by increasing its overall trade with the rest of the world.

In terms of wider economic dependence, the country has been consistently reliant on gold mining since independence from the USSR, the removal of Soviet subsidies, and the transition to a market economy. The country's reliance on Kumtor, a Canadian-owned gold mine accounting for approximately 8% of GDP (with worker remittances equating to approximately 28% of GDP in 2019) (World Bank, 2020b) has been well-documented, and not without controversy (European Commission, 2003; Satke, 2015; The Guardian, 2016; McGee, 2020). In 2013 it was reported that the Kyrgyz President in power, Almazbek Atambayev, was considering nationalising the economically vital Kumtor mine following a long-running dispute between joint venture partners, the Canadian based company Centerra and the Kyrgyz government, over mine ownership (BBC News, 2013).

Prior to 1991, under the central planning decisions of the USSR, Kyrgyzstan's production specialisations included the raising of sheep and production of wool, the cultivation of low quality grades of cotton, and the production of electrical energy in hydroelectric plants and of certain components for the defence industry (Dabrowski et al., 1995). Thompson and Foster (2003: 171–172) have documented that:

> Prior to the disintegration of the Soviet Union, Kyrgyzstan was already among the poorest of the Soviet regions with a gross national product per capita of US$1550 in 1991 compared with US$2470 in Kazakhstan (Anderson, 1999). The creation of the new Kyrgyz Republic, combined with the state of emergency declared in 1990 as a result of fighting on the Kyrgyz–Uzbek border, led to the onset of economic crisis in Kyrgyzstan. The level of inflation jumped from 200% in 1991 to 900% in 1992 (Anderson, 1999). As economic reform progressed and external investment and assistance increased, tourism was identified as an important industry sector and potential means of attracting revenue from the developed world.

Selm and Wagener (1993) report that Kyrgyzstan's share of foreign trade (i.e., trade with areas outside the republic's borders) represented 46% of GDP in 1988, comprised of 40% trade with other USSR republics and 6% trade with countries outside of the USSR. Hence, the country was heavily dependent on post-colonial trade relationships.

Examining post-colonial trade between Russia and the former Soviet republics, Mazhikeyev and Edwards (2020: 1) have reported that improvements in the global economy and international economic ties have contributed to "sharp recoveries in GDP levels for both Russia and many of the Central Asian Countries." However, Kyrgyzstan's loss of foreign markets during initial economic reform has been noted (Abazov, 1999; Dana, 2000) and it must be acknowledged that the country remains challenged by political instability. Indeed, Verdier (cited in Burgess, 2017) noted that Kyrgyzstan has had more than 20 presidents/prime ministers in 25 years and two revolutions. At the time of writing, in 2021, the latest Prime Minister, Sadyr Japarov, was appointed as head of government in Kyrgyzstan in November 2020, following a contentious parliamentary election (Putz, 2020).

Current events perhaps distract from the efforts of the country to develop positive international attention and integrate into the world market since independence (Guttman, 1999). It is notable that Kyrgyzstan was amongst the first former Soviet republics to implement economic reforms and transition towards a market-based economy (Price, 2018). It was also the first of the CIS countries to commit to programmes of international cooperation, joining UNESCO in 1992, the World Tourism Organization in 1993, and the World Trade Organization in 1998. A wide range of inter-regional tourism events have been hosted by the country and there is evidence of collaboration in tourism initiatives with countries from within and outside of Asia. Attempts at international engagement are apparent. As noted by Thompson and Foster (2003: 177),

> By early 1993, only 2 years after gaining independence from the Soviet Union, it was reported that Kyrgyzstan had been recognized by 120 nations and had established diplomatic relations with 61 of them.
>
> (The Library of Congress, 1996)

One of the key national development policy objectives introduced by President Askar Akayev, in office from 1990 until the 2005 Tulip Revolution (Olcott, 2005), focused on stimulating foreign investment to promote economic reform (Thompson and Foster, 2003). It is interesting to note that between 1997 and 2000, foreign investment worth approximately US$368 million was attracted to Kyrgyzstan; 43% of this was shared between banking, farming, trade, tourism, and services (Interfax, 2000). Post-Soviet independence, Spector (2008: 164) has acknowledged that the country continued to operate through a highly centralised regime, describing "a system in which the president holds significant formal political power and control over economic assets" and businesses were able to "receive preferential customs or tax treatment…not via legislation, but by gaining access to top customs or tax officials who can informally arrange deals." Mateeva (2010) has argued that there are few signs of stabilisation in the political trajectory of the state.

Kyrgyzstan's Tourism Context

Despite continual changes in political leadership and periods of instability, tourism has been continually identified as a national development priority. There have been seven national government tourism development strategies during independence but reductions in state budgets for tourism have affected policy implementation (Palmer, 2014).

Most recently, tourism has been included in the government strategy until 2040 as a top four priority sector of the economy (Fingar, 2018) and travel and tourism accounts for approximately 3.8% of GDP, showing a tendency to fluctuate but, overall, decrease between 2000 and 2019 (Knoema, 2020).

Tourism in Kyrgyzstan predates Soviet rule and post-communist independence. There are many historical accounts of travels across Central Asia, heavily influenced by the Silk Road network of trading routes running between Rome in Italy and China before the Ottoman Empire boycotted trade with China.

In terms of modern tourism, the conditions under Soviet rule (1917–1991) have received academic attention. Werner (2003) has documented the development of tourism in Central Asia, noting the strict control of foreign tourists during the Soviet period, particularly those from capitalist countries. She has also identified a dearth of hotels designated for foreign tourists and the restrictions on Western tourist stays in Bishkek beyond one night due to the presence of Soviet military facilities. This highly regulated approach placed the state as a key tourism actor, operating the foreign tourism agency "Intourist" that operated tour packages and hotel management. The command economy provided little or no space for the influence of other tourism actors. Initial transition to a market economy saw a gradual, conservative move towards privatisation, favouring alliances with "known" countries in line with Lucke and Eisler's (2016) observations about familiarity and cultural proximity but also reflecting internal disintegration within the USSR.

Indeed, examining the Issyk-Kul region, Kyrgyzstan's most lucrative tourist destination in terms of all tourism income, Palmer (2009: 186) has noted:

> Post-Soviet privatisation of the economy has resulted in many resort accommodation establishments being bought by persons linked to the former body or organisation to which the establishments previously belonged (during the Soviet era). This has resulted in establishments belonging to organisations based in a particular post-Soviet republic being acquired by entities within that republic (Nusorov, 2001 as cited in Allen, 2006). For example, three of the largest north shore resort accommodation establishments - the Karaganda, Hotel Kazakhstan and the Royal Beach - are now owned and operated by corporations based in Kazakhstan.

Zhukov (2001) observes that, outside of the tourism sphere, foreign firms, largely firms located in the CIS, have purchased Kyrgyz state facilities considered to be "non-essential". To date, the main tourism investment has come from Russia (59.7% of FDI in Kyrgyz tourism).

The observations of Koenker (2003) and Gorsuch and Koenker (2006) highlight the importance of Soviet-era tourism as an integral part of socialist ideology and acknowledge mass domestic tourism as a deliberate strategy (Koenker, 2003). Sustainability was not an immediate consideration. Reporting in 2003, based on first-hand experience of the destination, Thompson and Foster (2003: 172) commented:

> Accommodation facilities in Bishkek dating from the Soviet era, and the sanitoria on the northern shore of Lake Ysyk-Köl, are badly designed and constructed and environmentally unsympathetic to their surroundings, while other areas remain entirely undeveloped.

However, a desire by the Kyrgyz Government to privatise existing tourism resources and encourage the development of new products and services to stimulate tourism growth has been apparent and there have been clear efforts to encourage both foreign investment and indigenous entrepreneurship within the process of transferring ownership from the state to the private sector in the tourism industry.

Turdumamketov (2014) has argued that a positive balance of international tourism in Kyrgyzstan may be identified as one of few "effective" forms of international economic activity. Baxtishodovich et al. (2017) have identified the important interest and potential of unique cultural, historical, archaeological, and natural attractions to attract international tourists. However, Claytor (cited in Jasek, 2005) has argued that foreign tourist businesses are challenged by bureaucracy, high taxes, bad roads, and a cumbersome visa process, impacting on the country's ability to develop as an international tourist destination.

FDI and TFDI in Kyrgyzstan

The trajectory of development in Kyrgyzstan has been heavily impacted on by the emergence and shift in political governance structures and institutions. Institutional influence, over time, may be identified as a key variable affecting the formulation and implementation of policy. This includes TFDI policy, primarily the concern of central government, aligned to broader fiscal planning.

The OECD (2017) has noted how the Central Asian republics started signing international investment agreements (IIAs), including BITs, right after they gained independence. The speed at which this materialised and the transitional nature of political and economic development impacted on the ability of the newly independent republics to provide sound investment climates. The OECD (2017) has noted that a common tactic of layering different investment provisions covering the same country relations has prompted the bringing of claims by private investors against host governments. Legal objections have been raised based on the realisation that some investors have secured more favourable treaties. FDI has not occurred without conflict and contestation.

Indeed, a report by Knottnerus and Satke (2017: 3), focusing on the Kyrgyz Republic's experience with investment treaties and arbitration cases, stated that:

> The Kyrgyz Republic currently faces investment claims in the proximity of 1 billion USD, arising out of investment dispute settlement cases. If these have to be paid out, this will have a serious impact on the public budget of a country where 32% of the population lives below the poverty line.

Many analyses of Kyrgyzstan draw on the politico-economic history of the country and the trajectory of development in terms of why particular decisions were made rather than specifying why particular paths were *not* taken or how pathways were affected or interrupted due "critical junctures" – "situations of uncertainty in which decisions of important actors are causally decisive for the selection of one path of institutional development over other possible paths" (Capoccia, 2016: 95). In the words of Zappettini and Krzyżanowski (2019: 382):

> junctures are regarded as 'critical' because they set in motion path-dependent processes – in other words self-reinforcing trajectories – that become difficult to reverse as they eventually consolidate into one specific dominant institutional setup. At the same time, in a critical juncture, the contingent context in which choices are debated and made can also be seen as the result of institutional, cultural, and political trajectories which are reliant on antecedent conditions.

The sequencing of change and the idea of development *options* is pertinent when examining the Central Asian context. The region has endured a chequered past, characterised by periods of dictatorship. The economic shock of post-Soviet independence in 1991 still endures – "The economic fallout of independence was significant. The economy collapsed with USSR, as almost all of its exports were for the Soviet Union" (Burgess, 2017: online). However, the trajectory of economic development for Kyrgyzstan differs from its CIS neighbours. Recent data on balance of payments for Kyrgyzstan (World Bank, 2020a) indicates a negative overall trade balance (−11.2) in 2019 in contrast to Kazakhstan (−3.6), Tajikistan (−2.3), and Uzbekistan (−5.8).

Since the break-up of the Soviet Union, Russia, China, and the USA have been noted to be in competition for political and military influence in Central Asia. Kyrgyzstan remains notably dependent on Russia, approaching 30 years since Soviet independence. It has been identified to be "Russia's most loyal ally in Central Asia" (Mikovich, 2020). Overall, international trading links are low. Yet, an economic relationship between Kyrgyzstan and Turkey, albeit asymmetric, has also been observed (Yüceer, 2014) and Turkey has been described as a key partner of Kyrgyzstan (Murzaeva, 2014).

Vela (2011) has explored an increase in Turkish trade in Kyrgyzstan post-independence and has noted that cultural bonds and historical ties appear to underpin the developments. These appear to facilitate the building of trust and create aversion to the reported risks. The historical ties between Kyrgyzstan and Turkey are interesting, reflecting, at least to some extent pre-communist heritage and perceptions of cultural proximity. As a modern-day host destination for TFDI, questions may be raised over the desire for Turkish tourism companies to pursue outward investment in Kyrgyzstan. Jenish (2017: 22), for example, remarks that, "Kyrgyzstan pales in comparison with Turkey, a major rival in attracting Russian and Kazakh tourists in the 'Sun, Sand and Sea' vacation market".

It is perhaps notable that one of the first bilateral investment treaties established post-Soviet independence was between Turkey and the Kyrgyz Republic (dated 28 April, 1992 – ICSID, 2020). This treaty was signed at the same time as BITs between Kyrgyzstan's Central Asian neighbours: Turkey-Kazakstan; Turkey-Tajikistan; Turkey-Uzbekistan. Wheeler (2013: 1) has acknowledged that "the Central Asian states maintain a special place in Turkish foreign policy given ethno-linguistic Turkic ties." By 2018, Invest China (2018) was reporting Kyrgyzstan to have bilateral investment treaties with the USA, Armenia, Azerbaijan, Belarus, China, Finland, France, Georgia, Germany, India, Indonesia, Iran, Kazakhstan, the Republic of Korea, Lithuania, Malaysia, Moldova, Mongolia, Pakistan, Sweden, Switzerland, Tajikistan, Turkey, the UK, Ukraine, and Uzbekistan. The country's openness to FDI is noted although its investment climate is described as characterised by "considerable risk" (Invest China, 2018). Notably, the country ratified its membership of the International Center for the Settlement of Investment Disputes (ICSID) in 1997 and joined the New York Convention on the Recognition and Enforcement of Foreign Arbitral Awards in 1997.

Kyrgyzstan has relied heavily on technical assistance and civil society-building support from NGOs since the early 1990s. By the early 2000s, the country had become noted for having the highest NGO density in Central Asia (Garbutt and Heap, 2002). In 2020, Kakeev (cited in Imanaliyeva, 2020: 1) stated that "According to the government, there are 26,000 NGOs in Kyrgyzstan." Satke (2017) has noted that their presence is not without contention arguing that Kyrgyzstan's dependence on foreign aid brings risks to the country's social-economic development.

Research Approach

Following on from the review of literature and research context, two research questions were identified, relating to gaining a deeper understanding of TFDI in Kyrgyzstan:

- Are there key influences that have affected the path of a TFDI over time in Kyrgyzstan, Central Asia?
- How might exploring an evolutionary approach to analyse the path of a TFDI over time inform a wider understanding of tourism development?

Taking a qualitative approach will enable a more critical insight to be gained into TFDI in the context of Central Asia, a region that has an interesting and eclectic development past. The approach will allow for the temporal dimension to be acknowledged, pertinent to the situation of Kyrgyzstan as a designated "economy in transition" (Brück and Esenaliev, 2018).

The focus is on the evolutionary path of one TFDI case where a lasting interest "by a resident enterprise in one economy (direct investor) in an enterprise (direct investment enterprise) that is resident in an economy other than that of the direct investor" (OECD, undated: 7) is recognised to be most apparent in Kyrgyzstan. The Pinara-Bishkek Hotel was one of the first foreign hotels to be established as a joint venture in the republic following the dissolution of the USSR in 1991. The researcher is familiar with the destination area from her visits to Kyrgyzstan in the mid- to late-1990s. This first-hand experience was deemed to hold value given the challenges of being able to locate a broad spectrum of case information from authenticated secondary sources in the context under investigation and UK academic ethics committee restrictions affecting the employment of mixed methods during the unprecedented events of 2020.

The research took place during the COVID-19 global pandemic. This restricted the researcher's ability to engage first-hand with research participants based outside of the UK in a secure environment. A decision was made to focus on secondary data analysis. Desk research was undertaken to identify and systematically collate secondary qualitative data from publicly available online sources, including academic articles/theses, news media, and IDO/NGO reports. University ethics approval was granted.

Bennett and Elman's (2010) focus on process-tracing rooted in a realist epistemology was identified to be of interest and relevance. Furthermore, their earlier ideas (Bennett and Elman, 2006: 250) were recognised to offer value to the research approach:

> Within-case process tracing has also been identified as advantageous in addressing the complexity of path-dependent explanations and critical junctures – as for example with the development of political regime types – and their constituent elements of causal possibility, contingency, closure, and constraint.

The data collected existed in the format of text. Adopting an evolutionary lens, key events in the TFDI case example alongside the institutional context were identified over a 25-year period (1992–2017), through documentary analysis. Process tracing is used as a single case research method to make within-case inferences about the presence/absence of causal mechanisms as part of a theory-building approach, linked to the research questions presented in this chapter. These causal mechanisms are conceptualised to act as a system of interlocking parts that transmits causal forces between X and Y (Beach, 2016). In this study, the system of interest is referred to as "institutional context". Conceived as a system, it is made up of parts. These parts are viewed as "entities" that engage in "activities" and it is the activities or actions that may be recognised to be influential as "producers of change", transmitting "causal forces" or "causal possibilities".

Thus, causality is viewed as a dynamic, interactive influence of causes upon outcomes and a mechanism concerns agency by which an effect may be produced (Hernes, 1998).

"Institutional context" is defined in this study as "a set of institutional characteristics such as rules, regulations, policies, cultural factors and so on" (Ates, 2021: 41). The case under consideration is the Pinara-Bishkek Hotel as an example of TFDI. "Key events" are defined as notable or remarkable actions in the case example.

Figure 5.1 provides a visual summary of the findings to assist the discussion of "the evolutionary path of the Pinara-Bishkek Hotel TFDI".

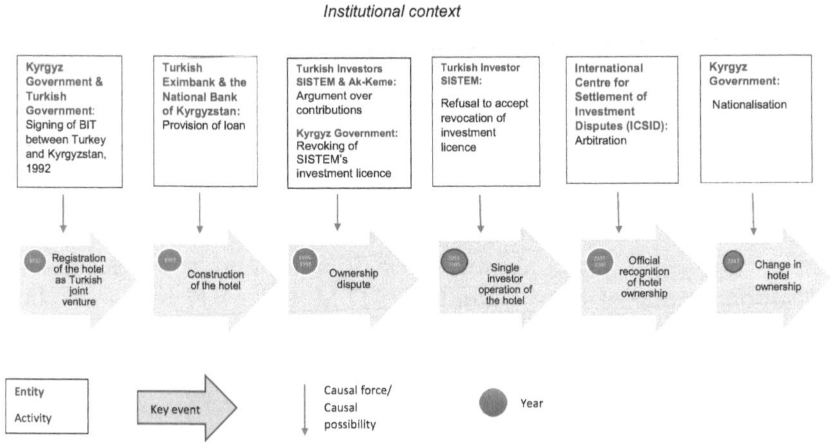

Figure 5.1 Process tracing the Pinara-Bishkek Hotel TFDI, Kyrgyzstan (1992–2017).

The Evolutionary Path of the Pinara-Bishkek Hotel TFDI

Each of the key events highlighted in Figure 5.1 is now examined and discussed, in turn, with consideration paid to the 'surrounding' institutional context and reference to literature presented earlier in this chapter.

Registration of the Hotel as a Turkish Joint Venture, 1992

The investment was established in Bishkek in 1992 through a joint venture between two Turkish entities, Sistem Mühendislik Insaat Sanayi ve Ticaret A.S. (SISTEM) and Ak-Keme (Jus Mundi, 2007). This followed on from the signing of a BIT between Turkey and Kyrgyzstan in 1992. The Turkish Government and the Kyrgyzstan Government as entities may be seen to have acted as 'producers of change' to produce a causal force that enabled the investment to be established. In terms of the wider institutional context changes in the regulatory context (instability and fragility within Kyrgyzstan – Mateeva, 2010), the policy context (market economy aspirations and a desire to forge international economic ties – Mazhikeyev and Edwards, 2020) and a legacy of Turkic-Kyrgyz cultural points of reference (Silk Road links during the Ottoman Empire and ethno-linguistic Turkic ties – Wheeler, 2013) may be recognised.

If we consider the assessment of the investment climate by the two Turkish companies, the potential for tourism development and the lack of facilities at an international standard (determinants of destination and investment attractiveness – Plog, 1974; Butler, 1980; Hu & Ritchie, 1993; Hu and Wall, 2005; Crouch, 2011; Andrades and Dimanche, 2017; Nazmfar et al., 2019; Kantarci, 2007) may also have influenced the perceived competitive environment and decision by the two Turkish investors to collaborate in an overseas hotel project. However, it is possible to view the establishment of the joint venture and the path taken at this stage – registration of the venture focused on the development of a hotel in Kyrgyzstan – as being enabled by institutional context as well as the actions of entities.

Construction of the Hotel, 1993

The construction of the hotel was assisted by the provision of a US $75 million loan from Turkish Eximbank to the National Bank of Kyrgyzstan. This enabled a credit line of US $6 million to be provided from the National Bank of Kyrgyzstan to SISTEM and Ak-Keme for the hotel project and may be recognised to be an action intended to support the stimulation of economic development in the Kyrgyz economy in line with expense calculations in submitted feasibility studies provided by the Turkish hotel investors to the Kyrgyz Ministry of Economics and Finances, State Committee on Foreign Economic Affairs, and the National Bank of Kyrgyzstan.

Here, we can note that the initial financing of the hotel development was fundamentally supported by a loan from public sector funds held by one state-owned national bank to another (Turkey to Kyrgyzstan). Construction of the hotel by the private investors was enabled through the two national banks and the two consecutive actions of providing a loan and providing a credit line. The influence of the state banks as entities and their actions on the path of the TFDI – enabling progress towards the build of the hotel – is apparent. The institutional context is perhaps less clear. The cultural context and historical ties between Turkey and Kyrgyzstan during a "New Great Game" of geostrategy and resource-competition may explain the action (Vela, 2011).

At this stage, the path to hotel construction may be acknowledged to be essentially caused by the availability of the finance (linked to the partnering of Turkey and Kyrgyzstan – Murzaeva, 2014) and the control given to a Kyrgyz state agency (operating through a highly centralised regime – Spector, 2008) to award the finance to the investor and directly manage the credit-debt financing process.

Ownership Dispute, 1995–1998

The Pinara-Bishkek Hotel was opened in August 1995 (Structurae, undated). It gained significance in the 1990s as the location for international political meetings. In 1999, the Indian newspaper *The Tribune* reported that the hotel had been the venue for a one-day political summit between Russian President Boris Yeltsin and Chinese leader Jiang Zemin focused on increasing stability along China's border with Russia and three Central Asian nations (The Tribune, 1999). However, the stability of the investment was threatened by conflict between the two Turkish investors.

Jus Mundi (2007) reports how, in December 1995, according to SISTEM, "Ak-Keme's armed men invaded the Hotel" and Ak-Keme informed SISTEM that it would no longer take responsibility for the safety of the lives of SISTEM's Turkish employees. A dispute between the two Turkish investors was apparent. It centred on the relative contributions of the two investors and culminated in the revoking of SISTEM's investment licence by the Kyrgyz government whilst Ak-Keme began to negotiate with alternative investors.

The original "Ak-Keme Pinara Joint Kyrgyz-Turkish Enterprise" was liquidated in 1998. SISTEM argued that the bankruptcy arose from Ak-Keme's failure to pay its debts, including debts to the Government of the Kyrgyz Republic, based on the loans from the Turkish Eximbank.

A new joint venture, "the Joint Kyrgyz Malaysian Venture" (JKMV), was created by Ak-Keme and a Malaysian investor ("Biznes Fokas Sdn. Bhd.") and ownership of the hotel was in dispute.

The actions of the Kyrgyz Government and the two Turkish investors, SISTEM and Ak-Keme, as entities may be seen to be a causal force in the outcome of the dispute – continued conflict – at this stage. Competing interpretations of the legal situation of the investment are apparent. The extent to which these reflect different levels of experience of foreign investment relations and the relative inexperience of Kyrgyzstan in working within the rules of an international treaty might be questioned. Mateeva (2010) has identified Kyrgyzstan in crisis, suffering from protracted periods of instability and a fragile state. From the perspective of the investors, the appeal of investing in countries that are "more corrupt and politically unstable compared to home" (Lucke and Eisler, 2016: 935) as a means of acquiring power and influence (Dwyer, 2014) may hold relevance.

Single Investor Operation of the Hotel, 2002–2005

SISTEM continued to operate the Pinara-Bishkek Hotel under a new registered enterprise, "Pinara Bishkek Ltd". SISTEM rejected the interpretation of the Kyrgyz Government, arguing that, despite revocation of its investment licence, many official documents recognised it as an investor. The investor occupied and operated the hotel from 1999 until 2005

This operation was not without difficulties. There were allegations from the Kyrgyz authorities that the construction of the hotel was defective and that it lacked the necessary licence. However, eventually in 2002 the Kyrgyz State Approval Committee officially approved the hotel for operation and the assets and debts of the original joint venture were formally transferred to SISTEM.

This action was however over-turned driven by a change in political climate in Kyrgyzstan (instability that culminated in the 2005 Tulip Revolution), shifting the attitude towards foreign investors. The Legislative Assembly of the Republic of Kyrgyzstan determined that hotel ownership should be transferred to a new company formed by Ak-Keme and a

Malaysian investor, JKMV. SISTEM continued to ignore these decisions and in 2005, following an armed takeover of the hotel by men claiming to be the Public Prosecutor and the General and police acting on behalf of Ak-Keme, the Kyrgyz Acting President and Prime Minister became involved, requesting that the hotel be returned to SISTEM. However, the hotel was not returned.

SISTEM appealed to the Government of the Kyrgyz Republic outlining the dispute and recalling the terms of the Main Agreement and the dispute settlement provisions of the BIT and requesting that employees of the JKMV be removed from the hotel. But the response was that the dispute lay between SISTEM and non-governmental people. SISTEM referred the case to ICSID for dispute arbitration.

These events draw attention to the role that conflict and contestation (Mahoney and Thelan, 2009; Conran and Thelen, 2016) play in macro-level TFDI agreements at a national level. Once again, the actions of the Kyrgyz Government and the two Turkish investors, SISTEM and Ak-Keme, as entities may be seen to be a causal force in the outcome of the dispute at this stage. Berdikulova and Soave (2020) note that the armed takeover of the hotel was possible because law enforcement in Kyrgyzstan was weakened during the 2005 revolution.

The agency and enacted power of one of the Turkish investors, SISTEM, in relation to continuing to pursue the development path of the hotel may be highlighted. Here, the work of Lucke and Eisler (2016) in relation to the agency of foreign investors may be seen to hold relevance alongside Dwyer's (2014) analysis of power and influence between TFDI "players".

Official Recognition of Hotel Ownership, 2007–2009

The referral of the ownership dispute by SISTEM to the ICSID was prompted by the Kyrgyz Government's view that it was not part of the dispute "between non-governmental people".

The arbitration tribunal decision was, however, categorical that the dispute arose directly out of SISTEM's investment and identified the Kyrgyz Republic, represented by its Government, a party to the dispute in its failure in its duty to protect SISTEM's investment under the Turkey-Kyrgyzstan BIT and determined that the Kyrgyz Government's actions were "tantamount to expropriation" (Invest China, 2018: 1). ICSID concluded that SISTEM's ownership of the hotel should be officially recognised and SISTEM should be provided with financial compensation from the Kyrgyz Government.

The actions of ICSID as an entity concluded the ownership dispute relating to the TFDI in favour of SISTEM. The failure of the Kyrgyz Government to act in accordance with the BIT supports Spector's (2013) observations of conflicts over securing property in Kyrgyzstan and other post-Soviet contexts, reflecting authoritarian regimes. Reports that the Kyrgyz Government has failed to provide compensation to SISTEM (Invest China, 2018;

U.S. Department of State, 2020) draw attention to resistance to international laws and the bypassing of legislation (Spector, 2013).

Indeed, despite the decision of the ICSID, by 2014 reports were circulating that the Kyrgyz courts had endorsed Ak-Keme's ownership of the hotel, ignoring SISTEM's ownership as determined by ICSID. It is difficult to trace this process but, looking at Trip Advisor reviews, by 2010 it is evident that the hotel had been re-named Ak Keme Hotel, Bishkek. SISTEM's involvement in the investment had terminated but the existence of the hotel as a TFDI had not ended.

Transfer to State Ownership, 2017

In 2017, it was reported that, following a contended independent appraisal of the value of the hotel, ownership of the property had been transferred to the state (Levina, 2017). Mainstream media news reports acknowledge that the "seizure" of the hotel from the President of the hotel Mr. R. Sarymsakov, noted as the original representative of Ak-Keme, had created animosity between the investor and the Kyrgyz Government. The investor claimed that alternative suggestions about the future of the hotel were presented to the Kyrgyz Prime Minister. However, Kostenko (2017: 1) has reported *that* "police officers cordoned off the hotel complex on June 22 early in the morning. All the employees of the hotel were taken out of the building."

The Ak-Keme Hotel remains listed at the original address of the Pinara-Bishkek Hotel and is advertised as open for bookings in 2021, indicating that it is still in operation as a 4-star, 170-room hotel.

The Kyrgyz Government as an entity may be recognised to have enacted a change in ownership of the hotel. The action may be seen to reflect authoritarian rule and is reminiscent of the former Soviet political regime of the country pre-independence.

Reflections on the Evolutionary Path of the Pinara-Bishkek Hotel TFDI

It may be identified that there are key influences that have affected the path of a TFDI over time in Kyrgyzstan, Central Asia. In particular, the relationship between the two Turkish investors who collaborated in the joint venture appears to be directly influential on the sequencing of change and the disruption of the development of the hotel as initially planned.

Additionally, the actions of the Kyrgyz Government must be recognised to have impacted upon the selection of one development path over another. Its responses to the dispute between the two investors highlight the entity as a key producer of change. The agency of the investors involved in the TFDI is apparent and it is evident that there has been resistance to the causal forces transmitted by the Kyrgyz Government. SISTEM's refusal to cease operating the hotel is a clear example of this.

The continuation of the hotel as a TFDI despite a protracted period of conflict is interesting. Although the hotel continues to operate, "closure" as a critical juncture may be recognised to have occurred when state ownership of the hotel ended its status as a TFDI. It is perhaps remarkable that "closure" does not appear to have been a decision taken by the investors or an outcome prior to 2017.

The issue of sovereignty as a risk attached to the signing of investment treaties (Knottnerus and Satke, 2017) appears to be relevant to the Pinara-Bishkek Hotel TFDI example. Reversion to nationalisation may be observed to be a typical control strategy employed by the Kyrgyz state when foreign investor relations break down.

Conclusions

The chapter set out to explore TFDI in the context of a transitional, post-Soviet economy through an evolutionary lens. It has demonstrated the scope to gain insights into both host government investment economy and foreign investor behaviours and actions over a sustained time period. This is important in terms of being able to

> "question the nature of that evolution with, for example, growth-oriented models often challenged as unsustainable in tourism".
>
> (Brouder and Eriksson, 2016: 384)

It is important to reflect on how the application of an evolutionary approach to explore the path of a TFDI over time might inform wider understanding of tourism development.

From a host country perspective, Khoshnevis et al.'s (2017) acknowledgement of the presence of hope and fear surrounding TFDI and its offerings is worthy of consideration. The relevance of contingency and risk associated with TFDI in Kyrgyzstan may be recognised through the case example of the Pinara-Bishkek Hotel, Kyrgyzstan. Issues of governance, political stability, and institutional capacities (Price, 2018) are highlighted as being highly influential for the development trajectory of a TFDI alongside collaboration and negotiation processes. This observation might inform wider understanding of tourism development by reinforcing Slocum and Backman's (2011) identification of the importance of "good" governance in achieving sustainability objectives alongside relationship skills. This is particularly pertinent in the context of less developed and transitional economies where increasing trade with the rest of the world is seen to be a pre-requisite of achieving sustainability goals (Allayarov et al., 2018) yet the historical legacy of centralised, authoritarian regimes has restricted relationship-building capacity. In the case of Kyrgyzstan, the speed at which foreign investment has occurred post-independence has restricted the ability of the state to develop "necessary policy space to harness investment to serve sustainable development" (Knottnerus and Satke, 2017). This indicates that the introduction of FDI and TFDI in transitional economies does not solely depend on opportunities but requires capability analysis, foresight, and planning.

The value of exploring an evolutionary approach to investigating TFDI has been highlighted. It has assisted identification of issues pertinent to the lasting interest (OECD, undated) of a TFDI in Kyrgyzstan and it has provided a fuller understanding of tourism processes underpinning development (Mellon and Bramwell, 2018). The approach has facilitated consideration of both structure and agency in tourism investment.

However, the difficulty in tracing the process beyond the internationally-reported arbitration tribunal decision in 2009 highlights the limitations of relying solely on secondary data analysis and indicates the potential for further research to gain a deeper understanding of how and why "specific paths unfold in particular ways over time and space" (Pike et al., 2016: 131).

References

Abazov, R., 1999. Economic migration in post-Soviet Central Asia: The case of Kyrgyzstan. *Post-Communist Economies*, *11*(2), pp. 237–252.

Allayarov, P., Mehmed, B., Arefin, S. and Nurmatov, N., 2018. The factors affecting Kyrgyzstan's bilateral trade: A gravity-model approach. *The Journal of Asian Finance, Economics and Business (JAFEB)*, *5*(4), pp. 95–100.

Ahn, M.J. and McKercher, B., 2015. The effect of cultural distance on tourism: A study of international visitors to Hong Kong. *Asia Pacific Journal of Tourism Research*, *20*(1), pp. 94–113.

AKIpress, 2012. *Building of Ak Keme Hotel under arrest, court processes may last several years – Finance Minister*. Kyrgyzstan, Business, July 17, 2012. URL: https://akipress.com/news:506461/

Al-Hallaq, S.S., Athamneh, A.B. and Suleiman, H.M., 2020. The impact of foreign direct investment on the growth of the tourism sector in Jordan (1980–2016). *Journal of Public Affairs*, *20*(2), p. e2005.

Andrades, L. and Dimanche, F., 2017. Destination competitiveness and tourism development in Russia: Issues and challenges. *Tourism Management*, *62*, pp. 360–376.

Ates, S., 2021. Corporate social performance and governance quality across the BRICS countries. In Yasser, Q. R. and Al-Mamun, A. (eds) *Transforming Corporate Governance and Developing Models for Board Effectiveness*. Pennsylvania: Premier Reference Source, IGI Global, pp. 1–343.

Baidoo, F.R., 2018. Welcome the Strangers! Tourism Development, Foreign Direct Investments and Economic Growth in Sub-Saharan Africa (Doctoral dissertation, University of Ghana).

Barrowclough, D., 2007. Foreign investment in tourism and small island developing states. *Tourism Economics*, *13*(4), pp. 615–638.

Baxtishodovich, B.S., Suyunovich, T.I. and Kholiqulov, A., 2017. The start-up of tourism in Central Asia Case of Uzbekistan. *World Scientific News*, *2*(67), pp. 219–237.

BBC News, 2013. Kyrgyzstan President: Key Goldmine may be Nationalised. November 7, 2013. URL: https://www.bbc.co.uk/news/world-asia-24852753.

Beach, D., 2016. It's all about mechanisms–what process-tracing case studies should be tracing. *New Political Economy*, *21*(5), pp. 463–472.

Bennett, A. and Elman, C., 2006. Complex Causal Relations and Case Study Methods: The Example of Path Dependence, *Political Analysis*, *14*(3), 250–267.

Bennett, A. and Elman, C., 2010. Case Study Methods. In Reus-Smit, C. and Snidal, D. (eds.) *The Oxford Handbook of International Relations* (pp. 499–517). Oxford: Oxford University Press.

Berdikulova, M. and Soave, T., 2020. Kyrgyz investment law and the problem of expropriation: Would the Indian Model BIT help? Budapest: Central European University.

Bianchi, R.V., 2003. Place and power in tourism development: Tracing the complex articulations of community and locality. *PASOS Revista de Turismo y Patrimonio Cultural*, *1*(1), pp. 13–32.

Boschma, R. and Martin, R., 2010. *The Aims and Scope of Evolutionary Economic Geography* (No. 1001). Utrecht University, Department of Human Geography and Spatial Planning, Group Economic Geography.

Britton, S.G., 1982. The political economy of tourism in the Third World. *Annals of Tourism Research*, *9*(3), pp. 331–358.

Brouder, P. and Eriksson, R.H., 2013. Tourism evolution: On the synergies of tourism studies and evolutionary economic geography. *Annals of Tourism Research*, *43*, pp. 370–389.

Bruck, T. and Esenaliev, D., 2017. Post-socialist transition and intergenerational educational mobility in Kyrgyzstan. *Economics of Transition and Institutional Change*, *26*(1), pp. 61–89.

Burgess, A., 2017. Life in Kyrgyzstan, 26 years after the collapse of the Soviet Union. Atlas Obscura, 30 August 2017. URL: https://www.atlasobscura.com/articles/kyrgyzstan-future-photographs-ussr#:~:text=On%20August%2031%2C%201991%2C%20as, Verdier%2C%20in%20an%20email%20interview.

Butler, R.W., 1980. The concept of a tourist area cycle of evolution: Implications for management of resources. *Canadian Geographer/Le Géographe canadien*, *24*(1), pp. 5–12.

Capoccia, G., 2016. Critical Junctures. In O. Fioretos, T. Falleti, and A. Sheingate (eds.), *The Oxford Handbook of Historical Institutionalism* (pp. 89–106). Oxford: Oxford University Press.

CEIC, undated. Kyrgyzstan foreign direct investment: Non CIS states: Turkey 1995–2017. National Statistical Committee of the Kyrgyz Republic. CEIC Data. URL: https://www.ceicdata.com/en/kyrgyzstan/foreign-direct-investment-by-country/foreign-direct-investment-non-cis-states-turkey

Conran, J. and Thelen, K.A., 2016. Institutional change. In Fioretos, T., Falleti, G. and Sheingate, A.D. (eds.) *The Oxford Handbook of Historical Institutionalism* (pp. 51–70). Oxford: Oxford University Press.

Copeland, B.R., 1991. Tourism, welfare and de-industrialization in a small open economy. *Economica*, *58*(232), pp. 515–529.

Crouch, G.I., 2011. Destination competitiveness: An analysis of determinant attributes. *Journal of Travel Research*, *50*(1), pp. 27–45.

Dabrowski, M., Jermakowicz, W., Pańków, J., Kloc, K. and Antczak, R., 1995. Economic reforms in Kyrgyzstan. *Communist Economies and Economic Transformation*, *7*(3), pp. 269–297.

Dana, L.P., 2000. Change and circumstance in Kyrgyz markets. *Qualitative Market Research: An International Journal*, *3*(2), pp. 62–73.

Deng, T., Hu, Y. and Yang, Y., 2019. How geographic, cultural, and institutional distances shape location choices of China's OFDI in tourism?–an empirical study on B&R countries. *Asia Pacific Journal of Tourism Research*, *24*(8), pp. 735–749.

Dexel, B., 2002. The illegal trade in snow leopards – A global perspective. German Society for Nature Conservation (NABU), October 2002. URL: https://snowleopardnetwork.org/bibliography/Dexel_2002.pdf

Dincer, H., Hacioglu, U., Tatoglu, E. and Delen, D., 2016. A fuzzy-hybrid analytic model to assess investors' perceptions for industry selection. *Decision Support Systems*, *86*, pp. 24–34.

Dirksmeier, P., 2017. Regionalentwicklung in einem postkolonialen Inselstaat im Globalen Süden: Eine empirische Analyse der Republik Mauritius mit Zensus-und Surveydaten. *Zeitschrift für Wirtschaftsgeographie*, *61*(3–4), pp. 119–134.

Donaldson, R. and Forssman, A., 2020 'Opening up to the world': An exploration of residents' opinions on and perceptions of St Helena island's tourism development. *African Journal of Hospitality, Tourism and Leisure*, *9*(6), pp. 944–958.

Drope, J., Chavez, J.J., Lencucha, R. and McGrady, B., 2014. The political economy of foreign direct investment—Evidence from the Philippines. *Policy and Society*, *33*(1), pp. 39–52.

Dunning, J. H., and Mcqueen, M., 1981. The eclectic theory of international production: A case study of the international hotel industry. *Managerial and Decision Economics*, *2*(4), 197–210.

Dwyer, L., 2014. Transnational corporations and the globalization of tourism. *The Wiley Blackwell Companion to Tourism* (pp. 197–209).

Edwards, M., Hulme, D. and Wallace, T., 1999. NGOs in a global future: Marrying local delivery to worldwide leverage. *Public Administration and Development: The International Journal of Management Research and Practice*, *19*(2), pp. 117–136.

Elhawary, S., Foresti, M. and Pantuliano, S., 2010, February. Development, security and transitions in fragile states. In *Meeting Series Report. Overseas Development Institute.*

Endo, K., 2006. Foreign direct investment in tourism—flows and volumes. *Tourism Management*, *27*(4), pp. 600–614.

Eurasianet, 2002. Kyrgyzstan. Traffic Accident Prompts Scrutiny of US Military Role in Kyrgyzstan, Mar 25, 2002. URL: https://eurasianet.org/traffic-accident-prompts-scrutiny-of-us-military-role-in-kyrgyzstan

European Commission, 2003. Cyanide Spill in Kyrgyzstan, Case Studies CHEST Project. URL: http://ec.europa.eu/health/ph_projects/2003/action3/docs/2003_3_09_cs4_en.pdf

Fauzel, S., Seetanah, B. and Sannassee, R.V., 2017. Analysing the impact of tourism foreign direct investment on economic growth: Evidence from a small island developing state. *Tourism Economics*, *23*(5), pp. 1042–1055.

Fingar, C., 2018 Kyrgyzstan's deputy tourism minister dreams big. FDI Intelligence, 15 February 2018. URL: https://www.fdiintelligence.com/article/70647

Garbutt, A. and Heap, S. (2002). *Growing Civil Society in Central Asia.* INTRAC Occasional Papers Series no. 39.

Global Security, undated. Bishkek. URL: https://www.globalsecurity.org/military/world/centralasia/kyrgyz-bishkek.htm

Gorsuch, A.E. and Koenker, D.P., 2006. *Turizm. The Russian and East European Tourist under Capitalism and Socialism.* Ithaca, NY/London: Cornell UP.

Green Economy Kyrgyzstan, 2018. *Green economy forum: Perspectives and progress of green economy development in the Kyrgyz Republic.* Ministry of Economy of the Kyrgyz Republic in cooperation with the Deutsche Gesellschaft für Internationale Zusammenarbeit (GIZ) GmbH, the UN Initiative Partnership for Action on Green Economy (PAGE) and BizExpert, October 19, 2018.

Guttman, C. (1999) *Kyrgyzstan: Breaking Out of the Old Shell.* UNESCO Courier, November, 21–22.

Henderson, J.C., 2003. The politics of tourism in Myanmar. *Current Issues in Tourism*, *6*(2), pp. 97–118.

Henderson, J.C., 2011. Tourism development and politics in the Philippines. *Tourismos*, *6*(2), pp. 159–173.

Hernes, G. (1998) 'Real virtuality', In P. Hedström and R. Swedberg (eds.), Social Mechanisms: An Analytical Approach to Social Theory, Cambridge: Cambridge University Press.

Hotel Online, 2021. Starwood Hotels and SERKA Turizm A.S. Open the Newly Built 240-room Sheraton Adana in Turkey. Sheraton Hotels & Resorts, September 4, 2014. URL: https://www.hotel-online.com/press_releases/release/starwood-hotels-and-serka-turizm-a-s-open-the-newly-built-240-room-sherato/

Hu, Y. and Ritchie, J.B., 1993. Measuring destination attractiveness: A contextual approach. *Journal of Travel Research*, *32*(2), pp. 25–34.

Hu, W. and Wall, G., 2005. Environmental management, environmental image and the competitive tourist attraction. *Journal of Sustainable Tourism*, *13*(6), pp. 617–635.

ICSID, 2020. *Sistem Muhendislik Insaat Sanayi ve Ticaret A.S. v. Kyrgyz Republic (ICSID Case No. ARB(AF)/06/1).* International Centre for Settlement of Investment Disputes. World Bank Group. URL: https://icsid.worldbank.org/cases/case-database/case-detail?CaseNo=ARB%28AF%29/06/1

Imanaliyeva, A., 2020. Kyrgyzstan: Draft bill threatens to drive NGOs against the wall. Eurasianet, May 22, 2020. URL: https://eurasianet.org/kyrgyzstan-draft-bill-threatens-to-drive-ngos-against-the-wall

International Business Council, 2019. Innovation for Sustainable Development Review of the Kyrgyz Republic Presented. IBC, February 28, 2019. URL: http://www.ibc.kg/en/news/ibc/4110_innovation_for_sustainable_development_review_of_the_kyrgyz_republic_presented

Interfax (2000) Interfax (2000) Kyrgyzstan attracts more direct investors in 2001. *The Times of Central Asia*, December 14, 2000.

Jasek, T., 2005. *Kyrgyzstan: Tourism Waits for its Moment to Shine. RadioFreeEurope/Radio Liberty*, October 4, 2005.

Jenish, N., 2017. *Tourism Sector in Kyrgyzstan: Trends and Challenges.* University of Central Asia. Graduate School of Development. Bishkek: Institute of Public Policy and Administration. URL: https://ucentralasia.org/Content/Downloads/UCA-IPPA-WP42_Eng.pdf

Jus Mundi, 2007. Sistem V. Kyrgyzstan. Sistem Mühendislik Inşaat Sanayi ve Ticaret A.Ş. v. Kyrgyz Republic, ICSID Case No. ARB(AF)/06/1. URL: https://jusmundi.com/en/document/decision/en-sistem-muhendislik-insaat-sanayi-ve-ticaret-a-s-v-kyrgyz-republic-decision-on-jurisdiction-thursday-13th-september-2007

Kantarci, K., 2007. Perceptions of foreign investors on the tourism market in central Asia including Kyrgyzstan, Kazakhstan, Uzbekistan, Turkmenistan. *Tourism Management*, *28*(3), pp. 820–829.

Keene, E. 2012. *Turkish Company Takes Aim at Kyrgyzstan's Centerra Shares*. November 7, 2012. Al-farabi news. Carnergie Endowment for International Peace. URL: https://carnegieendowment.org/2012/11/07/turkish-company-takes-aim-at-kyrgyzstan-s-centerra-shares-pub-49935

Khoshnevis Yazdi, S., Nateghian, N. and Sheikh Rezaie, N., 2017. The causality relationships between tourism development and foreign direct investment: An empirical study in EU countries. *Journal of Policy Research in Tourism, Leisure and Events*, *9*(3), pp. 247–263.

Knoema, 2020. Kyrgyzstan - Contribution of travel and tourism to GDP as a share of GDP. *World Data Atlas*. URL: https://knoema.com/atlas/Kyrgyzstan/topics/Tourism/Travel-and-Tourism-Total-Contribution-to-GDP/Contribution-of-travel-and-tourism-to-GDP-percent-of-GDP

Knottnerus, R. and Satke, R., 2017. *Kyrgyz Republic's Experience with Investment Treaties and Arbitration Cases*. Transnational Institute Amsterdam (July).

Koenker, D.P., 2003. Travel to work, travel to play: On Russian tourism, travel, and leisure. *Slavic Review*, *62*(4), pp. 657–665.

Kostenko, J., 2017. Ak-Keme hotel management tells about violations by government. 24 kg News Agency, June 28, 2017. URL: https://24.kg/english/56100_Ak-Keme_hotel_management_tells_about_violations_by_government/

Kozhokulov, S., Chen, X., Yang, D., Issanova, G., Samarkhanov, K. and Aliyeva, S., 2019. Assessment of tourism impact on the socio-economic spheres of the Issyk-Kul region (Kyrgyzstan). *Sustainability*, *11*(14), p. 3886.

Lee, C.K., Kang, S., Reisinger, Y. and Kim, N., 2012. Incongruence in destination image: Central Asia region. *Tourism Geographies*, *14*(4), pp. 599–624.

Levina, M. 2017. Bishkek's Ak-Keme hotel in Kyrgyzstan transferred to state ownership with help of police. *The Times of Central Asia*, 24 June 24, 2017. URL: https://www.timesca.com/index.php/news/18240-bishkek-s-ak-keme-hotel-in-kyrgyzstan-transferred-to-state-ownership-with-help-of-police

Litvin, S.W., Crotts, J.C. and Hefner, F.L., 2004. Cross-cultural tourist behaviour: A replication and extension involving Hofstede's uncertainty avoidance dimension. *International Journal of Tourism Research*, *6*(1), pp. 29–37.

Long, D.E., 1997. The impact of foreign investment on indigenous culture: An intellectual property perspective. *NCJ Int'l L. & Com. Reg.*, *23*, p. 229.

Lucke, N. and Eichler, S., 2016. Foreign direct investment: The role of institutional and cultural determinants. *Applied Economics*, *48*(11), pp. 935–956.

Ma, M., 2013. Tourism area development from an evolutionary economic geography perspective—The case of Guilin, China (Doctoral dissertation, Christian-Albrechts Universität Kiel).

Mahoney, J. and Thelen, K. (eds.), 2009. *Explaining Institutional Change: Ambiguity, Agency, and Power*. New York: Cambridge University Press.

Matveeva, A., 2010. *Kyrgyzstan in Crisis: Permanent Revolution and the Curse of Nationalism*. London: Crisis States Research Centre.

Mazhikeyev, A. and Edwards, T.H., 2020. Post-colonial trade between Russia and former Soviet republics: Back to big brother? *Economic Change and Restructuring*, *93*(1) pp. 1–42.

McGee, N., 2020. Questions raised about Centerra Gold's safety practices after another death at Kumtor mine. *The Globe and Mail*, February 18, 2020. URL: https://www.theglobeandmail.com/business/article-questions-raised-about-centerra-golds-safety-practices-after-another/

Mellon, V. and Bramwell, B., 2018. The temporal evolution of tourism institutions. *Annals of Tourism Research*, *69*, pp. 42–52.

Meyer, D., 2013. Exploring the duality of structure and agency–the changing dependency paradigms of tourism development on the Swahili coast of Kenya and Zanzibar. *Current Issues in Tourism*, *16*(7–8), pp. 773–791.

Mikovic, N., 2020. Kyrgyzstan's turmoil and the competition for Central Asia. *The Interpreter*, October 15, 2020. URL: https://www.lowyinstitute.org/the-interpreter/kyrgyzstan-s-turmoil-and-competition-central-asia

Mitlin, D., Hickey, S. and Bebbington, A., 2007. Reclaiming development? NGOs and the challenge of alternatives. *World Development*, *35*(10), pp. 1699–1720.

Murzaeva, D., 2014. Kyrgyzstan-Turkey relations: Cooperation in political and educational spheres. *Review of European Studies*, *6*, p. 39.

Nazmfar, H., Eshghei, A., Alavi, S. and Pourmoradian, S., 2019. Analysis of travel and tourism competitiveness index in middle-east countries. *Asia Pacific Journal of Tourism Research*, *24*(6), pp. 501–513.

Nørgaard, O., 2000. *Economic Institutions and Democratic Reform: A Comparative Analysis of Post-Communist Countries*. Massachusetts, USA: Edward Elgar Publishing.

OECD, undated. Glossary of foreign direct investment terms and definitions. *OECD Benchmark Definition of Foreign Direct Investment*, 4th Edition. URL: https://www.oecd.org/daf/inv/investment-policy/2487495.pdf

OECD, 2017. Investment Policy Reviews: Kazakhstan. URL: https://www.oecd-ilibrary.org/finance-and-investment/oecd-investment-policy-reviews-kazakhstan-2017_9789264269606-en

Olcott, M., 2005. Kyrgyzstan's "Tulip Revolution". Carnegie Endowment for International Peace, March 25, 2005.

Palmer, N., 2009. Domestic holidaymaking in a Central Asian context: Tourism legacies from pre-communist and Soviet times-Lake Issyk-Kul, Kyrgyzstan. *Domestic tourism in Asia. Diversity and Divergence*, pp. 181–196.

Palmer N.J., 2014. Kyrgyzstan, tourism. In Jafari, J. and Xiao, H. (eds.) *Encyclopedia of Tourism*. Cham: Springer. https://doi.org/10.1007/978-3-319-01669-6_611-1

Parliamentary Assembly, 2003. *Globalisation and Sustainable Development Report*. London: Committee on the Environment, Agriculture and Local and Regional Affairs. URL: https://assembly.coe.int/nw/xml/XRef/X2H-Xref-ViewHTML.asp?FileID=9961&lang=EN

Perić, J. and Radić, N.M., 2011. Sustainable foreign direct investment in tourism sector of developing countries. *Tourism in South East Europe, May* (2011), pp. 263–278.

Pike, A., MacKinnon, D., Cumbers, A., Dawley, S. and McMaster, R., 2016. Doing evolution in economic geography. *Economic Geography*, *92*(2), pp. 123–144.

Plog, S.C., 1974. Why destination areas rise and fall in popularity. *Cornell Hotel and Restaurant Administration Quarterly*, *14*(4), pp. 55–58.

Price, R. (2018) *Economic Development in Kyrgyzstan*. Institute of Development Studies, 30 July 2018.

Putz, C., 2020. From Prison to Parliament: Meet Kyrgyzstan's New Prime Minister. *The Diplomat*, October 14, 2020. URL: https://thediplomat.com/2020/10/from-prison-to-parliament-meet-kyrgyzstans-new-prime-minister/

Quer, D., Claver, E. and Andreu, R., 2007. Foreign market entry mode in the hotel industry: The impact of country-and firm-specific factors. *International Business Review*, *16*(3), pp. 362–376.

RFE/RL, 2005. Turkish businessman alleges property seizure in Kyrgyzstan. Newsline - April 5, 2005. RadioFreeEurope/ RadioLiberty. URL: https://www.rferl.org/a/1143371.html

Sanford Jr, D.M. and Dong, H., 2000. Investment in familiar territory: Tourism and new foreign direct investment. *Tourism Economics*, *6*(3), pp. 205–219.

Satke, R., 2015. Kyrgyzstan's controversial gold mine. *The Diplomat*, February 19, 2015. URL: https://thediplomat.com/2015/02/kyrgystans-controversial-gold-mine/

Satke, R., 2017. The downside of foreign aid in Kyrgyzstan. *The Diplomat*, June 10, 2017. URL: https://thediplomat.com/2017/06/the-downside-of-foreign-aid-in-kyrgyzstan/

Savina, L., 2014. Kyrgyzstan's business community calls on the country's leaders to resolve international disputes by negotiations. *IBC News*. International Business Council, November 18, 2014. URL: http://www.ibc.kg/en/news/ibc/529_kyrgyzstans_business_community_calls_on_the_countrys_leaders_to_resolve_international_disputes_by_negotiations

Selm, G.V. and Wagene, H.R., 1993. Former Soviet Republics' Economic Interdependence, *Osteuropa Wirtschaft*, 1, pp. 1–17.

Simon, H.A., 1955. On a class of skew distribution functions. *Biometrika*, *42*, pp. 425–440.

Spector, R.A., 2008. Securing property in contemporary Kyrgyzstan. *Post-Soviet Affairs*, *24*(2), pp. 149–176.

Structurae, undated. *Pinara Bishkek Hotel. Data.* URL: https://structurae.net/en/structures/pinara-bishkek-hotel

The Guardian, 2016. Tensions flare over environmental threat of Canadian gold mine in Kyrgyzstan. January 5, 2016. URL: https://www.theguardian.com/global-development/2016/jan/05/environmental-threat-canadian-gold-mine-kyrgyzstan

The Tribune, 1999. World, Thursday, August 26, 1999. URL: https://www.tribuneindia.com/1999/99aug26/world.htm

Thompson, K. and Foster, N., 2003. Ecotourism development and government policy in Kyrgyzstan. In Fennell, D.A. and Dowling, R.K. (ed.s) *Ecotourism Policy and Planning* (pp. 169–186). Wallingford: CABI.

Tiwari, A., 2011. Tourism, Exports and FDI as a Means of Growth: Evidence from four Asian Countries. *Romanian Economic Journal*, *14*(40) pp. 131–151.

Turdumambetov, B., 2014. Tourism development in the post-soviet and post-revolutionary country: A case study of Kyrgyzstan. *International Journal of Liberal Arts and Social Science*, *2*(6), pp. 73–87.

UNCTAD, 2004. Prospects for FDI flows, TNC strategies and promotion policies: 2004–2007. TD(XI)/BP/5, 27 April, Geneva.

UNCTAD, 2010. *Promoting Foreign Investment in Tourism.* New York and Geneva: Author.

UNCTAD, 2016. *The Investment Policy Review (IPR) of Kyrgyzstan.* March 2016, United Nations.

UNCTAD, 2019. World Investment Report 2019. United Nations. URL: https://unctad.org/system/files/official-document/wir2019_overview_en.pdf

U.S. Department of State, 2020. Report overview. 2018 investment climate statements: Kyrgyz Republic. URL: https://www.state.gov/reports/2018-investment-climate-statements/kyrgyz-republic/

Vela, J. 2011. *Kyrgyzstan: Banking on Turkish Trade*, eurasianet, May 10, 2011. URL: https://eurasianet.org/kyrgyzstan-banking-on-turkish-trade

Werner, C., 2003. The new silk road: Mediators and tourism development in Central Asia. *Ethnology*, *42*(2), pp. 141–159.

Wheeler, T., 2013. Turkey's role and interests in Central Asia. *Saferworld Briefing*, *3*, pp. 1–16.

WikimediaCommons,2020.File:Kyrgyzstan.png.URL:https://commons.wikimedia.org/wiki/File:Kyrgyzstan.png

World Bank, 2020a. Current account balance (% of GDP) – Kyrgyz Republic. URL: https://data.worldbank.org/indicator/BN.CAB.XOKA.GD.ZS?locations=KG

World Bank, 2020b. *Kyrgyz Republic*. Washington DC: The World Bank Group.

Worldometer, 2021. Kyrgyzstan Population (live). URL: https://www.worldometers.info/world-population/kyrgyzstan-population/

Yüceer, D., 2014. Kyrgyzstan-Turkey Relations: Cooperation in Political and Educational Spheres. *Review of European Studies*, *6*(3), p. 39.

Zappettini, F. and Krzyżanowski, M., 2019. The critical juncture of Brexit in media & political discourses: From national-populist imaginary to cross-national social and political crisis. *Critical Discourse Studies*, *16*(4), pp. 381–388.

Zhukov, S., 2001. Central Asia: Development under conditions of globalization. *Global Economic Review*, *30*(2), pp. 3–24.

Part III

Destination Case Studies in Tourism Foreign Direct Investment

6 Panel Analysis of FDI Determinants in Tourism Industry Revisited

New Evidence from France

António Miguel Valente Martins and Susana Cró

Introduction

Traditionally, the tourism industry has received a lower priority of policymakers than the manufacturing or agriculture sector since it has not been recognized as an appropriate and significant source of growth (UNCTAD 2007). Progressively, nonetheless, the tourism sector is being valued for its contribution to economic growth, employment, poverty reduction, earning exports revenues, consumer demand, capital formation, tax income, promotion of economic diversification, and a more services-oriented economy, helping to stimulate urban areas and cultural activities in decline (Endo 2006; UNCTAD 2007). But given that the tourism industry needs capital (some tourism activities are relatively capital-intensive), knowledge, infrastructure, and access to global marketing and distribution chains, FDI is often considered the most effective way to access these critical success factors (UNCTAD 2007:6).

Notwithstanding the importance of FDI in the tourism industry, the dynamics of FDI and their impacts on tourism have been relatively little studied (Endo 2006; UNCTAD 2007). Endo (2006) and Song, Dwyer, Li, and Cao (2012) refer to the lack of comprehensive FDI data in the hospitality industry as the main inhibiting factor. Few studies have investigated the country-specific location determinants of FDI activity in the hospitality industry (e.g., Dunning and Kundu 1995; Kundu and Contractor 1999; Johnson and Vanetti 2005; UNCTAD 2007; Assaf, Josiassen and Agbola 2015; Kristjánsdóttir 2015; Falk 2016; Nikšić Radić 2018). The studies can be divided into two distinct lines of investigation. The most recent studies investigate the country-specific location determinants of hospitality FDI industry using regression analysis (e.g., Assaf et al. 2015; Kristjánsdóttir 2015; Falk 2016; Nikšić Radić 2018), while previous studies using mostly questionnaire surveys to know the opinion of hotel investors or hotel managers/owners about the main location factors (e.g., Dunning and Kundu 1995; Kundu and Contractor 1999; Johnson and Vanetti 2005; UNCTAD 2007).

DOI: 10.4324/9781003155492-9

As mentioned above, the compilation of standardized FDI statistics in tourism is practically impossible. France is perhaps the only country that provides publicly available bilateral stock FDI data in tourism segmented by industry and investor country. According to the French Ministry of Foreign Affairs and International Development, France is the world's top tourist destination, with 83 million foreign tourists and accounting for 7% of French GDP,[1] and is among the top 10 most attractive countries for FDI.[2] Nevertheless, the FDI in hotels and restaurants (that will be used by us as a proxy of FDI in tourism) represented only 2.5% of total FDI in France in 2014. The importance of the tourism industry in the country and the availability of FDI data in the tourism sector are the reasons why France has been selected by us.

A generalized panel gravity model is estimated to update the determinants of bilateral inward FDI stock in the French hospitality industry. Results show that bilateral inward FDI stocks between France and investor countries are positively affected by their income and are inversely proportional to the distance between them. It is also found that differences between France and the investing countries in terms of taxes, labour costs, the abundance of skilled labour, the supply of public goods, and total FDI stock also play a significant role in understanding the foreign location decisions. Finally, the results show that France is particularly successful in attracting FDI in the hospitality industry from French-speaking countries with a common border and cultural proximity with France.

The present research and our previous empirical study (Cró and Martins 2020) differ from previous studies in two ways. First, as Falk (2016) we use a gravity model to study the determinants of FDI activity in the hospitality industry. As mentioned by Morley, Rosselló, and Santana-Gallego (2014) the gravity models have the advantage of taking into account the source country characteristics and the geographical distance. But our study differs from Falk (2016) to the extent that we use French bilateral inward FDI stock in the hospitality industry disaggregated by 19 countries and Falk (2016) uses a database of 2,420 FDI projects in 104 host countries. On the other hand, we include in our analysis three different measures of tax rates, agglomeration economies, and supply of public goods variables which were not considered in the Falk (2016) model. Second, by focusing on France, we can provide new insights into the determinants of FDI stocks in the hospitality industry, given the special characteristics of the country—the world's top tourist destination and one of the most attractive countries for FDI.

The structure of the remaining chapter is as follows. Section "Literature Review" provides some conceptual background and a review of previous studies while section "Methodology for Estimating FDI in Tourism" provides the empirical model. Data are presented in section "Data", and section "Empirical Results" reports and discusses the results. Section "Conclusion" concludes the paper.

Literature Review

Eclectic Paradigm and Modes of Entry into Foreign Markets

The question of why a given country has been successful in attracting FDI can be answered regarding the Eclectic Paradigm or OLI Paradigm Theory (Dunning 1977, 1981; Dunning and McQueen 1981; Markusen 1995). The Eclectic Paradigm merges various theories (e.g. Trade Theory, Theory of Industrial Organization, and Theory of the Firm) and asserts that the firm's FDI activities will be determined by its advantages related to ownership, internationalization and location. Each of these theories are now explained in detail. Ownership advantages—among other factors, the size of the firm and its capacity to obtain economies of scale, strategic brand development and international experience, and level of technological development of the firm. Location-specific advantages—among other factors, the growth, size, and level of development of the overall tourism market in the host country, the number and type of attractions, general infrastructure for tourism, the policy of government towards FDI, the stability of the country, and the physical and cultural distance between the host and home countries. Internalization advantages is the most abstract of the three—among other factors, economies of interdependent activities, that given the diversity in the tourism and the difficulties in organizing intermediary product markets, it is a factor for hotel firms to internalize their operations (for more details see Dunning 1977, 1981; Dunning and McQueen 1981; Markusen 1995; Johnson and Vanetti 2005).

According to IMF and OECD definitions, direct investment

> reflects the aim of obtaining a lasting interest by a resident entity of one economy (direct investor) in an enterprise that is resident in another economy (the direct investment enterprise). The "lasting interest" implies the existence of a long-term relationship between the direct investor and the direct investment enterprise and a significant degree of influence on the management of the latter.

"A direct investment relationship is established when the direct investor has acquired 10% or more of the ordinary shares or voting power of a company abroad" (Duce 2003: 2–3). Therefore, FDI investment in a foreign country can take the form of mergers and acquisitions (sole-venture or joint-venture) or greenfield investment (Endo 2006; Falk 2016). However, the major type of foreign entry in the tourism industry is non-equity participation—leasing agreement, management contract, and franchise agreement or some form of the marketing agreement, instead of equity modes such as FDI (see for example Dunning and McQueen 1982; Contractor and Kundu 1998; Altinay 2005; Endo 2006). According to the literature, the entry modes into foreign markets vary across host country characteristics. In high-income countries, franchising is the preferred way of entry into foreign markets (Dunning and McQueen 1981). On the other hand, Contractor and Kundu (1998) argue that equity investment contracts are quite common in developing countries, whereas non-equity models such as management and franchising contracts are preferred in high-income countries. Finally, in risky markets or countries, the preferred way of entry are non-equity modes such as management and franchising contracts to the extent that ways of entry are commonly regarded as less risky models of the entry (Chen and Dimou 2005).

Determinant Factors of Hotel FDI Attraction Process

Dunning and McQueen (1981) based on the Eclectic Paradigm were the first to make a deep attempt to explain the forms of Transnational Corporations' (TNCs) involvement in the international lodging industry. The authors report three reasons for internationalization in the lodging industry related to the Eclectic Paradigm: Ownership, Location, and Internalization (OLI) Advantages. The purpose of the subsequent studies was to identify the determinant factors of FDI attraction in the tourism industry. But as noted by Song et al. (2012: 1669) the lack of comprehensive data has bewildered researchers. Consequently, the studies about FDI in the tourism sector remain quite low compared to FDI in other sectors. Endo (2006: 600) argues that a wide range of activities that tourism covers (e.g., transportation, restaurants and hotels, tour operators, tour guides, travel agents, marketing and supply of souvenirs, and financial services for tourists) make the compilation of standardized FDI statistics in tourism at an international level almost impossible. As mentioned by Endo (2006) surprisingly, the relevant data are scarce even among developed countries. Despite these limitations, Table 6.1 presents a summary of the main studies carried out within the hotel sector about determinant factors of the hotel FDI attraction process. Our study is particularly interested in country-specific location factors.

As can be seen from Table 6.1, the literature has identified several locational factors for FDI in the hotel industry. However, except Falk (2016), none of the mentioned studies have used bilateral FDI flow data to investigate the determinants of FDI flows in the hotel sector. Falk (2016), based on the fDi Markets database which contains a register of cross-border investment projects from around the world (2003 onwards), analysed the bilateral FDI flow of 2,417 FDI projects in the hotel industry related to the construction of new and expansion of existing hotels by an international investor. Given the limitations in obtaining data, we use a panel gravity model to study the bilateral inward FDI in tourism—using stock data on FDI in hotels and restaurants as a proxy—for France. France is perhaps the only country that provides FDI stock data publicly available segmented by industry (with data for the hotel and restaurant industry) and investor country, and for that reason has been selected. The FDI gravity model is explained in detail in the next section.

Table 6.1 Summary of location factors of hotel FDI attraction process

Author(s)	*Location factors for FDI*	*Additional information*
Dunning and Kundu (1995)	Market size Growth rate Tourism opportunities Availability of infrastructure Political and economic stability	Application of Eclectic Paradigm on the international Hotel industry Field study
Kundu and Contractor (1999)	Market size proxied by GDP and tourism revenues has the main location factors Other explanatory variables: Population Ratio of exports to GDP Country FDI rating Total inward FDI into nation	Data on FDI flow in the hotel sector for 67 host countries Examines location choices for equity investment Questionnaire survey for global hotel chains
Johnson and Vanetti (2005)	Home-country proximity Market size and growth Infrastructure and tourist attractions Perception of region reputation Government incentives	Questionnaire survey for leading chains, framed around an eclectic paradigm Study for hotel operators in five countries in eastern central Europe
Endo (2006)	Historical, cultural, and geographical distance Level of economic development Cost-based factors (taxation and labour costs) Political and/or economic risks Industry privatisation and FDI regulation Investment incentives Socioeconomic environment Infrastructures quality	The author analyses the impact of FDI on host countries using inward FDI data Determinants of FDI in tourism are virtually the same as the other sectors
UNCTAD (2007)	Tourism demand from developed countries Economic growth Market size FDI incentives FDI related regulation Geographical and cultural proximity	Questionnaire survey to identify the host-country determinants of FDI in the hotel industry
Snyman and Saayman (2009)	Health and safety (crime, HIV/AIDS, malaria) Political stability Cost factors and skills Infrastructure (roads, airports) Market size (international tourism demand, GDP) Tourism-specific amenities and assets (e.g., beaches, cultural sites, and natural environment)	Questionnaire Survey send to estate agents in South Africa that specialise in dealing with foreign direct investors List of 42 host-country characteristics that may affect FDI in tourism industry
Steiner (2010)	Business regulations and host-market growth with a highest influence in tourism FDI than violent political unrest	Time-series data for Egypt

(*Continued*)

Author(s)	*Location factors for FDI*	*Additional information*
Guillet et al. (2011)	Government policies Presence of local entrepreneurs Mega events Market potential	Data from secondary sources using keyword research Outlook on the future of the investments proposed by multinational hotel corporations in China
Zhang et al. (2012)	Market size and demand measured by total inbound tourists and average inbound tourist spending Actual FDI Business environment measured by GDP per capita Policy and mega-events	Time-series data for Chinese provinces The authors analyse the factors that determine the location strategies of multinational hotel groups into Chinese provinces
Assaf et al. (2015)	Most important factors: Welcomeness Infrastructure Socioeconomic factors (crime rate and corruption) Other factors: Opportunities for tourism Quality of human resources Political stability Restrictions and regulations Cultural and development proximity Price advantage	Random effects panel data model for more than 120 host countries for the period 2007–2011 The authors identify 23 factors that matter most for international hotels when selecting host destinations The authors make a ranking of these factors
Kristjánsdóttir (2015)	Economic and market size Taxes Skilled labour Cultural distance	OLS estimations for the determinants of FDI inflows in the hospitality industry into the OECD countries with special emphasis on Nordic countries
Falk (2016)	Market size Common language Business regulations Tax rates Minimum wages	The author study the determinants of hospitality FDI industry using a Gravity Model Database of 2,420 FDI projects carried out by 50 parent countries in 104 host countries from 2005 to 2011
Li et al. (2017)	Investment environment Outbound tourism scale to host country Tourism economy scale Trade level Innovation capability	The authors investigate the factors that influence China's outward foreign direct investment in tourism They employ a panel dataset involving 21 host countries for 10 years (2004–2013)
Nikšić Radić (2018)	Terrorism GDP growth International tourist arrivals Political stability Corruption Business start ease index	The author studies the determinants of FDI in Tourism Industry for a list of 50 countries in the period 2000–2016 The study uses system-GMM estimator for dynamic panel data models

Methodology for Estimating FDI in Tourism

The Gravity Model

The gravity model applies Newton's universal law of gravitation to study the trade or economic flows between countries and assumes that bilateral flows between any two countries are positively affected by their income (mass) and inversely proportional to the distance between them. Developed during the 1960s, with the pioneering studies of Tinberg (1962) and Linneman (1966), the gravity models were criticized due to this lack of theoretical background for the gravity equations (in this respect see Fratianni, Marchionne, and Oh 2011; Morley, Rosselló, and Santana-Gallego 2014). For this reason, "the gravity models were neglected in the tourism literature during the eighties and the nineties. An illustration of this abandonment can be found in the fact that surveys on tourism demand do not explicitly cite gravity models" (Morley et al. 2014: 2). However, recent studies explore the validity of the gravity model to explain tourism flows, and the empirical evidence supports the applicability and robustness of traditional gravity factors to the flow of trade and FDI flows and stocks in general (Keum 2010; Blonigen and Piger 2014).

The basic gravity model can be expressed in the following logarithmic form:

$$\ln F_{IJ} = \beta + \alpha \ln GDP_I + \lambda \ln GDP_J + \xi \ln Dist_{IJ} + \varepsilon_{IJ} \tag{6.1}$$

where $\boldsymbol{F_{IJ}}$ is the international flow between the country I and J; ***GDP*** denotes the gross domestic product of each country; ***Dist*** refers to the distance between the country I and J; ε_{IJ} is a normal error term; and $\boldsymbol{\beta}$, $\boldsymbol{\alpha}$, $\boldsymbol{\lambda}$, and $\boldsymbol{\xi}$ are parameters to be estimated. The ***GDP*** captures the market dimension and given that large countries have a greater potential than small countries for investing abroad, a positive effect is expected for this variable. In turn, ***Dist*** (which is a proxy for transaction, transportation, or, more generally, information costs) is an important determinant of FDI flows. It is expected that FDI activity decreases with the geographical distance between pairs of countries. However, geographic distance also stands for information and transaction costs and cultural distance. Among other authors, Rodríguez (2002) and Ivanov and Ivanova (2016), demonstrate that hotel chain managers have a higher propensity to invest in countries whose cultural profile is closer to the cultural profile of its home country because they will be more familiar with the cultural environment for doing business there than in other countries.

Finally, in line with several studies using gravity models, Kogut and Singh's (1988) cultural distance index[3] and two dummies—common border and common language—are included in the analysis. These variables are introduced to catch cultural factors and information and transaction costs that significantly contribute to FDI activity linkages between countries, for instance through network externalities (e.g., Kristjánsdóttir 2015; Ivanov and Ivanova 2016; Falk 2016).

Extended Gravity Model

We also test the impact of taxation in tourism FDI activity by adding to our baseline model three different tax rates: (i) statutory corporate income tax (*STAT*); (ii) bilateral effective average tax rates on corporate income (*beatrs*) and (iii) bilateral effective marginal tax rate on corporate income (*bemtrs*). A majority of the studies use the statutory corporate income tax as a measure of the corporate tax burden on FDI (deMooij and Ederveen 2003).[4] From a conceptual and empirical point of view *beatrs* or *bemtrs* should be used, because effective tax rates reflect tax incentives correctly when compared to statutory taxes and are better candidates to measure the impact of taxation on FDI activity (Devereux and Griffith 2002, 2003; Devereux, Griffith, and Klemm 2002).

Some other empirical studies have raised the question of incorporating agglomeration economies jointly with taxes into the analysis of FDI flows or stocks. Wheeler and Mody (1992) show that agglomeration economies are an important determinant for US multinationals' investment-location decisions. They use three different measures of agglomeration economies—degree of industrialization, infrastructure quality, and existing stock of FDI—and find that agglomeration-related factors seem to clearly dominate the decision where to invest. They also show that infrastructure quality arises to be an especially important factor in developing countries. Bénassy-Quéré, Fontagné, and Lahréche-Révil (2005) in their study about investment-location decisions between 11 OECD countries find that their measure of agglomeration—market potential, together with taxes is an important determinant of bilateral FDI flows. Finally, Hansson and Olofsdotter (2013) study the impact of tax differentials and agglomeration economies on FDI. The authors use four agglomerations variables—market potential, host-country stock of FDI, country's population density, and expenditure on research and development (R&D)—and find that "the impact of taxes is sensitive to the estimation method and the inclusion of agglomeration economies and there is some evidence of agglomeration mitigating the negative effects of taxes on FDI flows" (Hansson and Olofsdotter 2013: 2654).

Moreover, Bénassy-Quéré et al. (2005) and Falk (2016), among other authors, show that low labour costs in the host country have traditionally been an important factor in the decision to invest in other countries. Falk (2016) argues that labour cost differentials between a pair of countries play an important role in hotel FDI, particularly for cost-saving vertical FDI. On the other hand, Kristjánsdóttir (2015) in her research of FDI in the hospitality industry in Iceland and Norway, in comparison to the Nordics and a range of other OECD countries, includes a measure for skilled labour abundance in the country. The empirical results show that skilled labour is a key factor to attract FDI into the hospitality sector.

Finally, as stated by Bénassy-Quéré et al. (2005) and Hansson and Olofsdotter (2013), countries have the possibility of compensating for high taxation by the provision of public goods. We use the government investment expenditure (in % of GDP) to measure the impact of public investment in FDI. We expect a positive effect on FDI, as this proxy is likely to be closer to the building of public goods (Bénassy-Quéré et al. 2005).

We follow previous studies on FDI, and we employ a gravity framework (see, e.g., Bénassy-Quéré et al. 2005; Hansson and Olofsdotter 2013) where inward FDI stock for the French hotel and restaurant industry is determined by standard gravity variables as well as taxes, labour costs, and skills, agglomeration economies, country-pair-specific effects, and a time dummy (Egger and Pfaffermayr 2003). The model applied is a generalized panel gravity model, with the following log-linear specification as our baseline estimation:

$$\ln FDI\ stock_{ijt} = \beta_1 \ln GDP_{it} + \beta_2 \ln GDP_{jt} + \beta_3 \ln Dist_{ijt} + \beta_4 X_{ijt} + \beta_5 Z_{ij} + \beta_6 W_{jt} + \gamma_t + \alpha_{ij} + \varepsilon_{ijt} \qquad (6.2)$$

where $\mathbf{ln}\boldsymbol{FDIstock}_{ijt}$, is the log of the stock of hotel and restaurant FDI from home country i to host country j in year t; $\mathbf{ln}\boldsymbol{GDP}_{it}$ and $\mathbf{ln}\boldsymbol{GDP}_{jt}$ are the log of the investing and host country's GDP, respectively; $\mathbf{ln}\boldsymbol{Dist}_{ijt}$ is the log of the distance between country i and j; $\boldsymbol{X}_{ijt}$ are location factors which vary between country-pairs and over time (taxes, skilled labour, labour costs, market potential variables); $\boldsymbol{Z}_{ij}$ are location factors which vary over country-pairs only (common border and language dummies and Kogut and Singh's (1988) cultural distance index); $\boldsymbol{W}_{jt}$ are location factors which vary over time and host countries (government expenditure); γ_t are time dummies; α_{ij} are country-pair-specific effects; and ε_{ijt} is the remainder error term.

Data

To study the key determinants of French inward FDI in the hospitality industry, we collect data on bilateral FDI stock between France other 19 countries[5] between 2000 and 2019. French inward FDI stock activity (in millions of EUR) data from the Central Bank of France are used, and the OECD industry classification of "Hotels and Restaurants" is applied as a proxy of FDI in tourism (an identical proxy was used by Endo 2006). We use inflow FDI stock data instead of flow data because the first one reflects better the long-term investment incentives of the firms (see, e.g., Endo 2006, in this regard). The total of French FDI stock in the Hotels and Restaurants industry in 2019 was 2,527 million EUR. European countries accounted for 61% of the total value followed by the US with 22%. In Europe, the UK, Netherlands, and Luxembourg, with 25%, 17%, and 11% of total FDI stock in the Hotels and Restaurants industry in 2019 are the main investor countries.

The GDP data in purchasing parity standards were obtained from the World Bank. Data on distance and dummies for countries with common language and border are drawn from Mayer and Zignago (2006). Distance is measured in kilometres between the principal cities of countries weighted by population size. In addition to geographical distance, the importance of cultural distance is taken into account. Cultural distance is calculated according to Kogut and Singh's (1988) formula that is based on Hofstede's five cultural dimensions: Power distance, masculinity, individualism, uncertainty tolerance, and long-term orientation.

Data for corporate tax differentials are calculated as simple differences between the tax rates in the host country (France) and the investing country. Three tax variables are used in the study: (i) Statutory tax rates, (ii) average effective tax rates, and (iii) marginal effective tax rates. All of them are taken from the Oxford University Centre for Business Taxation. We also test the effects of agglomeration variables on FDI decisions. We consider two proxies of agglomeration economies in our study. The first measure is the differential between host-country market potential and the investor. The variable is defined as the ratio of the GDP of the host country to its average internal distance. The second measure is the differential between total FDI stock in the host-country and investor country. As stated by Hansson and Olofsdotter (2013), it is expected that countries with a larger stock of FDI will have an advantage in attracting new investment. Data for total FDI stock are sourced from the World Bank.

Moreover, two measures are included relating to labour costs and skills. Labour costs data are taken from the OECD. Relative unit labour costs are computed as the difference in the (natural logarithm of) unit labour costs in the host country against the investor country. The data on abundance of skilled labour variable are obtained from the World Bank.

Finally, to explore the possible compensation of high taxation by the supply of public goods, we include in the analysis the difference in government investment expenditures (% of GDP) in the host country relatively to investor country, collected from the OECD. All of these variables and further data description are presented in Table 6.2.

Empirical Results

The empirical results are reported in Table 6.3. The results are based on 299 observations (data of 81 observations are missing) for the period 2000–2019. Six specifications are presented. The statutory tax rate on corporate income simple differences between the tax rates in France and the investing country as a proxy of tax rates is used along with the dummies common language and common border and cultural distance index in the first specification. In specifications (3) and (4) we use bilateral effective average tax rate simple differences and in (5) and (6) the bilateral effective marginal tax rate simple differences, as proxies of tax rates. We exclude the dummies common language and common border, in the specifications (2), (4), and (6) given the strong correlation of both variables with Kogut and Singh's (1988) cultural distance index.

Table 6.2 Data description

Variable	*Definition*	*Source*	*Mean*	*SD*
Inward FDI stock	French bilateral inward FDI stocks, in "Hotels and Restaurant" industry, million euro.	Central Bank of France (www.banque-france.fr/economie-et-statistiques/base-de-donnees/investissements-directs.html).	251.6	595.6
GDP	Gross domestic product at market prices, millions of PPS (Purchasing Power Standards)	World Bank	599,456	488,233
Distance	Bilateral distance in kilometres between the largest cities in country *i* and country *j*, weighted by population size.	CEPII (www.cepii.fr).	2,596	3,037
Common Language	Dummy that takes the value of one when countries use the same language (in this case French).	CEPII (www.cepii.fr).	0.202	0.403
Common Border	Dummy that takes the value of one when countries have a common border.	CEPII (www.cepii.fr).	0.399	0.491
Cultural Distance	Kogut and Singh's (1988) cultural distance index calculated from secondary data, using the five Hofstede' cultural dimensions.	Hofstede Centre. (http://geert-hofstede.com/national-culture.html). (Own calculations)	8.188	4.898
Statutory Tax Rates	Statutory tax rate on corporate income simple differences between the tax rates in the host country and the investing country.	Oxford University Centre for Business Taxation (www.sbs.ox.ac.uk/faculty-research/tax/publications/data). (Own calculations)	0.059	0.069
Effective Average Tax Rate (*beatrs*)	Net present value (NPV) of tax payments as share of NPV of total pre-tax income simple differences between the tax rates in the host country and the investing country. See Devereux and Griffith (2003).		0.029	0.060
Effective Marginal Tax Rate (*bemtrs*)	The proportional difference between the pre-tax and post-tax required rates of returns simple differences between the tax rates in the host country and the investing country. See Devereux *et al.* (2002).		0.006	0.058
Market Potential	For country *i*: $\sum_{k \neq i} GDP_k / Distance_{ik}$, million euro. We use the difference between host country and investor country.	World Bank, CEPII (own calculations)	1,296	1,467

(*Continued*)

Variable	*Definition*	*Source*	*Mean*	*SD*
Total FDI Stock	Total stock of FDI, million euro. We use the difference in the (natural logarithm of) total stock FDI in the host against the investor.	World Bank	682,366	955,657
Unit Labour Costs	Relative unit labour costs are computed as the difference in the (natural logarithm of) unit labour costs in the host against the investor.	OECD, (own calculations)	3,456	12,012
Skilled Labour	Skilled labour abundance in the investing country, relative to the host. The variable skilled labour is presented by the measure "School enrollment, tertiary (% gross)".	World Bank	3.666	18.012
Government Investment	Government investment expenditures as a share of GDP. We use the difference of government investment in the host against the investor.	OECD	25.542	5.555

This table present the definition, source and two statistical descriptive measures – mean and standard deviation (SD), of the variables included in the Gravity Equations.

The results show that differences between France and the 19 investor countries in terms of taxes, labour costs, the abundance of skilled labour, the supply of public goods, and total FDI stock play a significant role in understanding the foreign location decisions.

The three measures of tax rates used in the estimations show a negative and statistically significant impact on FDI stock as expected. Identical result was obtained for labour cost differential. These results are not surprising since positive differences between France and the foreign investors in terms of taxes or labour costs are a signal of lower competitiveness of France in terms of costs, with a consequent negative impact on the ability of FDI attraction. A similar result was obtained by Falk (2016: 229) who writes that "cost-based considerations such as wages and taxes are important for FDI activity in the hotel sector". In turn, Bénassy-Quéré et al. (2005) find a positive relationship between labour costs and FDI, which is most likely attributable to an omitted variable problem. The authors themselves refer that labour costs may account for the impact of skill differentials across host countries. Following Kristjánsdóttir (2015), we include a proxy of skilled labour abundance in the analysis. Like Kristjánsdóttir (2015) we find a positive and statistically significant result.

Hansson and Olofsdotter (2013) use four measures of agglomeration to investigate if agglomeration economies mitigate the downward spiral in the tax rate. Two of these agglomeration measures are market potential and the total stock of FDI. Both were used by us in this study. The authors argue that it is expected that countries with a larger market potential and total

Table 6.3 Determinants of French inward stock FDI in hotels and restaurant industry

	(1)		*(2)*		*(3)*		*(4)*		*(5)*		*(6)*	
Intercept	−1.725***	(−6.706)	−1.925***	(−6.008)	−1.884***	(−5.218)	−1.799***	(−5.691)	−1.923***	(−6.743)	−1.856***	(−5.204)
$LnGDP_{FRANCE}$	6.111***	(7.122)	5.987***	(5.456)	6.234***	(5.449)	6.123***	(5.545)	5.244***	(6.518)	6.200***	(5.453)
$LnGDP_{INVESTOR}$	1.011***	(4.444)	0.788***	(3.234)	1.056***	(5.001)	0.798***	(3.345)	1.345***	(5.665)	0.954***	(3.812)
LnDIST	−1.488***	(−6.245)	−1.323***	(−5.234)	−1.865***	(−6.952)	−1.701***	(−5.456)	−2.000***	(−6.222)	−1.855***	(−5.241)
COM_LANG	2.555***	(5.033)	–	–	2.567***	(5.335)	–	–	2.765***	(5.456)	–	–
COM_BORD	1.492***	(3.321)	–	–	1.754***	(4.759)	–	–	1.567***	(4.345)	–	–
CULT_DIST	0.020	(0.299)	0.050**	(2.023)	0.027	(0.699)	0.082**	(2.221)	0.014	(0.567)	0.061**	(2.201)
STAT_TAX_differential	−4.333***	(−2.999)	−5.210***	(−3.123)	–	–	–	–	–	–	–	–
BEATRS_differential	–	–	–	–	−8.234***	(−3.234)	−7.856***	(−2.888)	–	–	–	–
BEMTRS_differential	–	–	–	–	–	–	–	–	−8.456***	(−3.025)	−7.756**	(−2.343)
M_POT_differential	−0.081	(−0.500)	−0.245	(−1.567)	−0.199	(−1.345)	−0.281*	(−1.759)	−0.309*	(−1.869)	−0.389**	(−2.111)
LnFDI_Stock_differential	0.257***	(3.162)	0.199***	(2.978)	0.277***	(3.145)	0.287***	(3.084)	0.255***	(2.988)	0.265***	(2.999)
LnUN_LAB_differential	−0.301*	(−1.827)	−0.299**	(−2.220)	−0.433**	(−2.344)	−0.222*	(−1.711)	−0.369**	(−2.143)	−0.254*	(−1.823)
SK_LAB_differential	0.022***	(2.998)	0.036***	(4.236)	0.020***	(2.995)	0.035***	(3.694)	0.024***	(3.222)	0.036***	(4.345)
GOV_EXP_differential	0.219***	(5.745)	0.165***	(5.123)	0.188***	(5.567)	0.181***	(5.021)	0.219***	(6.226)	0.180***	(4.935)
TIME_DUM	−0.499**	(−2.111)	−0.298*	(−1.845)	−0.576**	(−2.134)	−0.388*	(−1.825)	−0.567**	(−2.289)	−0.345*	(−1.725)
Number Observations	299		299		299		299		299		299	
Adjusted R^2	0.700		0.632		0.701		0.634		0.730		0.633	
Hausman test	29.300	p = [0.001]	34.256	p = [0.000]	29.123	p = [0.001]	31.546	p = [0.000]	30.345	p = [0.000]	31.470	p = [0.000]
F test	31.994	p = [0.000]	27.021	p = [0.000]	33.567	p = [0.000]	26.900	p = [0.000]	31.846	p = [0.000]	26.740	p = [0.000]
Breusch-Pagan LM test	865.01	p = [0.000]	813.05	p = [0.000]	856.45	p = [0.000]	817.13	p = [0.000]	890.25	p = [0.000]	823.35	p = [0.000]

This table reports the estimation results of six regressions based on equation (6.2). The dependent variable is the log of French bilateral inward FDI stocks, in "Hotels and Restaurant" industry between 2000 and 2019 (**LnInw_Stock_FDI**). **LnGDP** is the log of gross domestic product at market prices, millions of PPS (Purchasing Power Standards); **LnDIST** is the log of bilateral distance in kilometres between the largest cities in France (country *i*) and country *j*, weighted by population size; **COM_LANG** is a dummy that takes the value of one when countries speak the same language (in this case French); **COM_BORD** is a dummy that takes the value of one when countries have a common border with France; **CULT_DIST** is measured by Kogut and Singh's (1988) cultural distance index calculated from secondary data, using the five Hofstede' cultural dimensions; **STAT_TAX_differential** is the statutory tax rate on corporate income simple differences between the tax rates in France and the investing country; ***BEATRS*_differential** is the bilateral effective average tax rate simple differences between the tax rates in France and the investing country; ***BEMTRS*_ differential** is the bilateral effective marginal tax rate simple differences between the tax rates in France and the investing country; **M_POT_differential** is the difference between market potential of France and the investing country; **LnFDI_Stock_differential** is the difference between total FDI stock of France and the investing country (in logarithms); **LnUN_LAB_differential** is the relative unit labour costs computed as the difference in the (natural logarithm of) unit labour costs in France against the investor; **SK_LAB_differential** measure the skilled labour abundance in France relatively to investor country; **GOV_EXP_differential** is the difference between government investment expenditures as a share of GDP of France relatively to investor country. **TIME_DUM** is a dummy that takes the value of one for the years of the financial crisis – years of 2007–2010. We use a generalised panel gravity model. *t*-statistics are presented in brackets. ***, ** and * denote statistical significance at 1%, 5% and 10%, respectively. *p* refers to *p*-values.

stock of FDI will have an advantage in attracting new investment. Our results show a significant and positive relationship for the total stock of FDI. On the other hand, like Hansson and Olofsdotter (2013), no significant results are found for market potential.

Furthermore, a positive and statistically significant effect is found for the measure of government investment expenditures. This result is in line with the literature on tax competition that underlines the possible compensation of high taxation by the provision of public goods (e.g., Bénassy-Quéré et al. 2005).

The results also show that France is particularly successful in attracting FDI in the hospitality industry from French-speaking countries with a common border and cultural proximity with France. Specifications (1), (3), and (5) do not support the conclusion that cultural proximity is an important factor of FDI attraction to the extent that this variable does not show statistical significance. This result may be due to the strong correlation of common border and common language dummies with Kogut and Singh's (1988) cultural distance index. As can be seen in the other specifications, when the two dummies are not included in the estimation the cultural distance measure shows a significant and positive signal as expected. This result seems to support the idea that foreign investors have a greater propensity to invest in countries whose cultural profile is closer to the cultural profile of their home country. Similar results were found by Rodríguez (2002) and Ivanov and Ivanova (2016).

Looking at the gravity factors, the results show that bilateral inward FDI stocks between France and investor countries are positively affected by their income and are inversely proportional to the distance between them. Falk (2016) argues that a positive signal exhibited by host country GDP indicates that market-seeking considerations are relevant for hotel FDI. With a larger market, there is a higher probability that TNCs will be able to recover the costs of their FDI. We also include the French country size as another market-related variable and as expected, find a significant and positive signal. The geographical distance, which is a proxy for the transaction, transportation, or more generally, information costs, shows a significant and negative signal as expected.

Finally, we use a time dummy that takes the value of one for the years 2007–2010, to study the impact of the financial crisis on bilateral inward FDI stock. The coefficients of the time dummy present a negative and significant signal. Given that in these years of widespread financial crisis, investors' confidence levels experienced a strong shock, it is normal to register a decrease in foreign investment levels.

Conclusion

Our study aims to investigate the determinants of French inward FDI stock in the hospitality industry. A panel gravity model is applied to bilateral inward FDI stock between France and 19 investor countries in the Hotels and Restaurants industry over 2000–2019.

We conclude that bilateral inward FDI stocks between France and investor countries are significantly and positively affected by their income and are inversely proportional to the distance between them. A key finding of the study is that France is particularly successful in attracting FDI in the hospitality industry from French-speaking countries with a common border and cultural proximity with France. Furthermore, taxes and labour costs have a considerable negative impact on bilateral inward FDI activity. This result shows that cost-based considerations such as labour costs and taxes are an important determinant factor of FDI activity. The total FDI stock and market potential differential, two measures of agglomeration economies, reveal different results. The first one presents a positive and significant signal while the second measure does not seem to affect FDI stock. Other factors such as differences between France and the investing countries in terms of abundance of skilled labour and supply of public goods and the financial crisis are also relevant for hospitality FDI activity. Overall, we find that the most significant determinants of FDI in the hospitality industry are virtually the same as those obtained for total FDI stock.

These results have two important implications for tourism policymakers. First, given that FDI activity showed to be sensitive to cost factors such as labour costs and taxes, it is crucial to control these costs for countries to remain competitive. Second, the results show the possible mitigation of adverse effects of high taxation by the provision of public goods. This may be an alternative solution for countries like France, which has a high tax burden to remain competitive, given that the country utilizes a part of its high taxes for public infrastructure.

While this paper is a step towards a better understanding of FDI stock determinants, there are several limitations to our analysis. First, it focuses solely on one country—France—and hence, we ought not to draw generalized conclusions for all countries. It will be important in the future to extend this study to more countries as soon as there is a compilation of standardized FDI statistics on tourism. Second, and related to the previous point, we use FDI in hotels and restaurants as a proxy of FDI in tourism. This solution is adopted among others by Endo (2006) and Falk (2016). However, it excludes other activities related to tourism such as transportation, tour operators, tour guides, travel agents, the supply of souvenirs and marketing, and financial services for tourists, which may lead to some biases in the results obtained. However, the alternative solution would be the lack of studies on FDI in the tourism sector. Third, given that non-equity participation is the major common mode of foreign entry into the tourism industry instead of equity modes (in which FDI is included), again our results could suffer from some type of bias. Finally, future studies should examine the evolution of FDI in the hospitality sector with the arrival of the COVID-19 pandemic and the harmful effects caused by this.

Notes

1 Please see http://www.diplomatie.gouv.fr/en/french-foreign-policy/economic-diplomacy-foreign-trade/facts-about-france/one-figure-one-fact/article/france-the-world-s-leading-tourist.

2 Please see http://www.diplomatie.gouv.fr/en/french-foreign-policy/economic-diplomacy-foreign-trade/facts-about-france/one-figure-one-fact/article/france-back-among-the-top-10-most.

3 For further details about cultural distance measures, including Kogut and Singh's (1988) cultural distance index, and its impact in the tourism industry, see for example Ng et al. (2007). The authors refer that Kogut and Singh's (1988) cultural distance index is the most popular way to measure cultural distance in which three-quarters of the studies used this approach to measure the cultural distance.

4 For more details about the definition and construction of the three tax rates see DeMooij and Ederveen (2003), page 677.

5 The countries included in the sample are Belgium, Brazil, Canada, China, Denmark, Germany, India, Ireland, Italy, Japan, Luxembourg, Netherlands, Russia, Spain, Sweden, Switzerland, Turkey, the UK, the USA, and the host country, France. These countries are the only countries included in the sample because are the only ones for which data on FDI stocks in the hospitality industry is available.

References

Altinay, L. (2005). Factors Influencing Entry Mode Choices: Empirical Findings from an International Hotel Organisation. *Journal of Hospitality & Leisure Marketing*, 12 (3): 5–28.

Assaf, A., Josiassen, A. and Agbola, F. (2015). Attracting International Hotels: Locational Factors that Matter Most. *Tourism Management*, 47: 329–340.

Bénassy-Quéré, A., Fontagné, L. and Lahréche-Révil, A. (2005). How Does FDI React to Corporate Taxation? *International Tax and Public Finance*, 12: 583–603.

Blonigen, B. and Piger, J. (2014). Determinants of Foreign Direct Investment. *Canadian Journal of Economics/Revue Canadienne d'Économique*, 47 (3): 775–812.

Chen, J. and Dimou, I. (2005). Expansion Strategy of International Hotel Firms. *Journal of Business Research*, 58 (12): 1730–1740.

Contractor, F. and Kundu, S. (1998). Modal Choice in a World of Alliances: Analyzing Organizational Forms in the International Hotel Sector. *Journal of International Business Studies*, 29 (2): 325–357.

Cró, S. and Martins, A. M. (2020). Foreign Direct Investment in the Tourism Sector: The Case of France. *Tourism Management Perspectives*, 33: 100614.

deMooij, R. and Ederveen, S. (2003). Taxation and Foreign Direct Investment: A Synthesis of Empirical Research. *International Tax and Public Finance*, 10: 673–693.

Devereux, M. and Griffith, R. (2002). The Impact of Corporate Taxation on the Location of Capital: A Review. *Swedish Economic Policy Review*, 9: 79–102.

Devereux, M. and Griffith, R. (2003). Evaluating Tax Policy for Location Decisions. *International Tax and Public Finance*, 10 (2): 107–126.

Devereux, M., Griffith, R. and Klemm, A. (2002). Corporate Income Tax Reforms and International Tax Competition. *Economic Policy*, 35: 449–496.

Duce, M. (2003). Definitions of Foreign Direct Investment (FDI). A Methodological Note. Banco de Espana. Available here: https://www.bis.org/publ/cgfs22bde3.pdf.

Dunning, J. (1977). Trade, Location of Economic Activity and MNE: A Search for an Eclectic Approach. In Ohlin, B., Hesselborn and Wijkman, P. (eds.) *The International Allocation of Economic Activity* (pp. 395–418). London: Macmillan.

Dunning, J. (1981). *International Production and the Multinational Enterprise*. London: George Allen and Unwin.

Dunning, J. and Kundu, S. (1995). The Internationalization of the Hotel Industry: Some New Findings from a Field Study. *MIR: Management International Review*, 35 (2): 101–133.

Dunning, J. and McQueen, M. (1981). The Eclectic Theory of International Production: A Case Study of the International Hotel Industry. *Managerial and Decision Economics*, 2 (4): 197–210.

Dunning, J. and McQueen, M. (1982). Multinational Corporations in the International Hotel Industry. *Annals of Tourism Research*, 9 (1): 69–90.

Egger, P. and Pfaffermayr, M. (2003). The Proper Panel Econometric Specification of the Gravity Equation: A Three-way Model with Bilateral Interaction Effects. *Empirical Economics*, 28: 571–580.

Endo, K. (2006). Foreign Direct Investment in Tourism – Flows and Volumes. *Tourism Management*, 27: 600–614.

Falk, M. (2016). A Gravity Model of Foreign Direct Investment in the Hospitality Industry. *Tourism Management*, 55: 225–237.

Fratianni, M., Marchionne, F. and Oh, C. (2011). A Commentary on the Gravity Equation in International Business Research. *The Multinational Business Review*, 19 (1): 36–46.

Guillet, B., Zhang, H. and Gao, B. (2011). Interpreting the Mind of Multinational Hotel Investors: Future Trends and Implications in China. *International Journal of Hospitality Management*, 30: 222–232.

Hansson, A. and Olofsdotter, K. (2013). FDI, Taxes and Agglomeration Economies in the EU15. *Applied Economics*, 45 (18): 2653–2664.

Ivanov, S. and Ivanova, M. (2016). Market Penetration of Hotel Chains: Does Culture Matter? *Annals of Tourism Research*, 57: 234–278.

Johnson, C. and Vanetti, M. (2005). Locational Strategies of International Hotel Chains. *Annals of Tourism Research*, 32 (4): 1077–1099.

Keum, K. (2010). Tourism Flows and Trade Theory: A Panel Data Analysis with the Gravity Model. *The Annals of Regional Science*, 44: 541–557.

Kogut, B. and Singh, H. (1988). The Effect of National Culture on the Choice of Entry Mode. *Journal of International Business Studies*, 19 (3): 411–432.

Kristjánsdóttir, H. (2015). Foreign Direct Investment in the Hospitality Industry in Iceland and Norway, in Comparison to the Nordics and a Range of Other OECD Countries. *Scandinavian Journal of Hospitality and Tourism*, 16 (4): 395–403.

Kundu, S. and Contractor, F. (1999). Country Location Choices of Service Multinationals: An Empirical Study of the International Hotel Sector. *Journal of International Management*, 5 (4): 299–317.

Li, X., Huang, S. and Song, C. (2017). China's Outward Foreign Direct Investment in Tourism. *Tourism Management*, 59: 1–17.

Linneman, H. (1966). *An Econometric Study of International Trade Flows*. Amsterdam: North-Holland.

Markusen, J. (1995). The Boundaries of Multinational Enterprises and the Theory of International Trade. *Journal of Economic Perspectives*, 9 (2): 169–189.

Morley, C., Rosselló, J. and Santana-Gallego, M. (2014). Gravity Models for Turism Demand: Theory and Use. *Annals of Tourism Research*, 48: 1–10.

Mayer, T. and Zignago, S. (2006). *Notes on CEPII's Distance Measures*. Available here: http://www.cepii.fr/distance/noticedist_en.pdf.

Nikšić Radić, M. (2018). Terrorism as a Determinant of Attracting FDI in Tourism: Panel Analysis. *Sustainability*, 10 (12): 4553.

Ng, S., Lee, J. and Soutar, G. (2007). Tourists' Intention to Visit a Country: The Impact of Cultural Distance. *Tourism Management*, 28: 1497–1506.

Rodríguez, A. (2002). Determining Factors in Entry Choice for International Expansion. The Case of the Spanish Hotel Industry. *Tourism Management*, 23: 597–607.

Snyman, J. and Saayman, M. (2009). Key Factors Influencing Foreign Direct Investment in the Tourism Industry in South Africa. *Tourism Review*, 64 (3): 49–58.

Song, H., Dwyer, L., Li, G. and Cao, G. (2012). Tourism Economics Research: A Review and Assessment. *Annals of Tourism Research*, 39 (3): 1653–1682.

Steiner, C. (2010). An Overestimated Relationship? Violent Political Unrest and Tourism Foreign Direct Investment in the Middle East. *International Journal of Tourism Research*, 12 (6): 726–738.

Tinberg, J. (1962). *Shaping the World Economy: Suggestions for an International Economic Policy*. New York: The Twentieth Century Fund.

United Nations Conference on Trade and Development (UNCTAD). (2007). *FDI in Tourism: the Development Dimension*. United Nations, New York and Geneva, 2007. Available here: https://unctad.org/system/files/official-document/iteiia20075_en.pdf.

Wheeler, D. and Mody, A. (1992). International Investment Location Decisions: The Case of US Firms. *Journal of International Economics*, 33: 57–76.

Zhang, H., Guillet, B. and Gao, W. (2012). What Determines Multinational Hotel Groups' Locational Investment Choice in China? *International Journal of Hospitality Management*, 31 (2): 350–359.

7 Why Croatia Still Needs FDI in Tourism? Insights and Evidence

Maja Nikšić Radić

Introduction

The role tourism plays in the Croatian economy is incomparable with any other part of its economy. Given the fact that it generates as much as a quarter of its GDP, that role is almost intimidating. Of course, fear is not unreasonable when there is an awareness of the very nature of tourism. Tourism is by nature very volatile, sensitive to numerous external stimuli… it recovers quickly, but also stumbles quickly. It may suffice to recall its reaction to the global financial crisis of 2008, the sporadic terrorist attacks that hit many European metropolises during the second decade of the 21st century, the situation with COVID-19 that hit the world in early 2020 and still strongly affects the overall world economy, and immeasurably on tourism itself. On the other hand, tourism is part of the benefits of globalization and its importance will only grow in the future. It is well known that, according to the UNWTO (2015), the number of worldwide international tourist arrivals will increase by an average 3.3% a year from 2010 to 2030 and reach the level of 1.8 billion international tourist arrivals by 2030. As for Croatia, and quoting Orsini and Ostojić (2018) regarding the role of tourism in the Croatian economy, tourism will "remain a key sector of the Croatian economy."

What Croatia imposed as a challenge (of course with the necessity of strengthening the power of other parts of the economy) is to realize the greatest possible benefits from such a strong role of tourism. The current model of Croatian tourism development has certain weaknesses that affect the possibility of achieving even greater benefits. High seasonality, limited range of services and low average consumption are its most common weaknesses (Ministry of Tourism, 2013) whose reason certainly lies in the structure of Croatian accommodation capacities that significantly deviate from those at the EU level. The second challenge, and no less important, is to ensure the sustainability of Croatian tourism development. This sustainability came into question in 2019 when Croatia and Dubrovnik appeared on the World's Top 3 for Overtourism per Capita Map (Vivid Maps, 2019).

Additionally, attraction of FDI in Croatian tourism, especially hotels, has a significant opportunity to correct the Croatian structure of

DOI: 10.4324/9781003155492-10

accommodation capacities. Although the issue of new hotels may provoke debate, given the fact Croatia faced over-tourism in the summer months before the COVID-19 situation, limiting the number of hotels and similar accommodation could further aggravate the situation with over-tourism. According to Vorkinek (2020), restricting hotels and similar accommodation increases the number of accommodation facilities on Airbnb. Considering that over-tourism is closely connected with high seasonality, which is a major problem of Croatian tourism, the presence of global hotel brands would help to correct it. Hotel accommodation in Croatia has a much lower seasonality compared to the dominant private accommodation.

It has been proven that there is a positive relationship between the number of transnational corporations in tourism and the level of tourism revenues in the country (UNCTAD, 2007). Croatia has attracted a relatively modest level of FDI in tourism so far and still cannot boast of the presence of some global hotel brands such as Hyatt. Given the above, Croatia could gain additional financial benefits by attracting additional foreign capital to tourism. Previous research indicates how FDI in tourism has affected the productivity of tourism, the number of international tourism arrivals, but also the overall level of FDI in Croatia (Nikšić Radić, 2014; Perić and Nikšić Radić, 2016a; Bezić and Nikšić Radić, 2017). Finally, global hotel brands show greater resilience to sudden market instability, which would be of multiple benefits to Croatia, as a country living off tourism.

After the *Introduction*, the second section *Positioning of tourism in the Croatian economy* indicates the importance of tourism in the Croatian economy, but also its need to further attract foreign capital. The section on *Theoretical considerations – What are the benefits from FDI in tourism?* shows a theoretically grounded background of potential positive impacts that may result in the presence of foreign capital in tourism with an emphasis on hotels. The third section *FDI in Croatian economy and Croatian tourism* shows the current presence of foreign capital in the Croatian economy with special emphasis on the presence in tourism. *Achieved effects of FDI in Croatian tourism* is the title of the fourth section, which evaluates the previous positive effects that foreign capital has brought to Croatian tourism. The fifth section, *Croatia still needs FDI in tourism but is it attractive to foreign investors?* assesses the potential attractiveness of Croatian tourism to further attract foreign capital.

Positioning of Tourism in the Croatian Economy

Croatia is a country that is historically oriented towards tourism. The heritage of tourism dates back to the time of the Habsburg Monarchy. Spas for the wealthy elite emerged in the late 18th century, and flourished in the 19th century at a time when the development of transport (railways and steamships) connected the Viennese metropolis with the outlying territories (Allcock, 1991). The tradition of organized tourism in Croatia is about 150 years long, although even before that, at the beginning of the 19th century, there were phenomena similar to tourism (such as going on pilgrimages or treatment), so the first inns, lodgings, hotels and spas were built for this purpose. It is interesting to emphasize that the first beginnings of tourism in Croatia are associated with the continental part of the country. It was not until the end of the 19th century that tourism began to be linked to the coastal part of the country (Gržinić, 2019).

The "Hygienic Society" founded in 1869 on Hvar and large hotels such as "Kvarner" in Opatija and "Imperial" in Dubrovnik founded in the next three decades enhanced the reach of this early form of health tourism (Pirjevec, 1998). The tourists continued to come in these former Habsburg resorts during the first Jugoslav state and The Ministry of Trade and Industry has established a Department for the Traffic of Foreign Visitors (Ballinger, 2003). After World War II tourism in Croatia is increasingly gaining importance, and achieved strong momentum with an average of one million tourist arrivals per year (Vukonić, 2005). The JUG Tourist Association, founded in 1927, also sought to promote winter tourism in places that were previously outside tourist circles (Vasilevska, 1998).

Croatian tourism was already significantly developed while it was still part of the former Yugoslavia and accounted for over 90% of the Yugoslav Adriatic coastline. Croatia, like ex-Yugoslavia, recorded a special boom in tourism in the period from the 1960s to the 1980s. Yugoslavia focused on attracting tourists from Western Europe (Cattaruzza and Sintès, 2012). In 1985, Croatia had 41 million or 82% of all overnight stays by foreign tourists in Yugoslavia (Gosar, 1989). The relative openness to the West enabled the arrival of European tourists and led to a form of mass tourism (Grandits and Taylor, 2010). After the disintegration of Yugoslavia and the end of the Homeland War, Croatia very quickly successfully branded itself on the foreign market and became a recognizable tourist destination.

A relevant indicator of the development of the tourism sector of a particular country is the number of international tourist arrivals. The number of international tourist arrivals in Croatia in the period from 1965 to 2020 is shown in Figure 7.1.

Yet as a member of ex-Yugoslavia, Croatia recorded strong growth in international tourist arrivals, when already in 1985 could boast of more than ten million international tourist arrivals. With the disintegration of Yugoslavia and the beginning of the Homeland War, tourism in Croatia lost its contours until the repeated establishment of security. Croatian tourism reached such a level from the last century in 2010, when the number of international tourist arrivals exceeded a record ten million international tourist arrivals for the second time in its history. The global financial crisis briefly stopped the expansion of international tourist arrivals, but Croatian tourism continued to grow at a gallop, reaching more than 19 million international tourist arrivals in 2019 and questioning the sustainability of Croatian tourism. With the onset of the pandemic in 2020, Croatian tourism achieved, given the circumstances, a satisfactory seven million tourist arrivals.

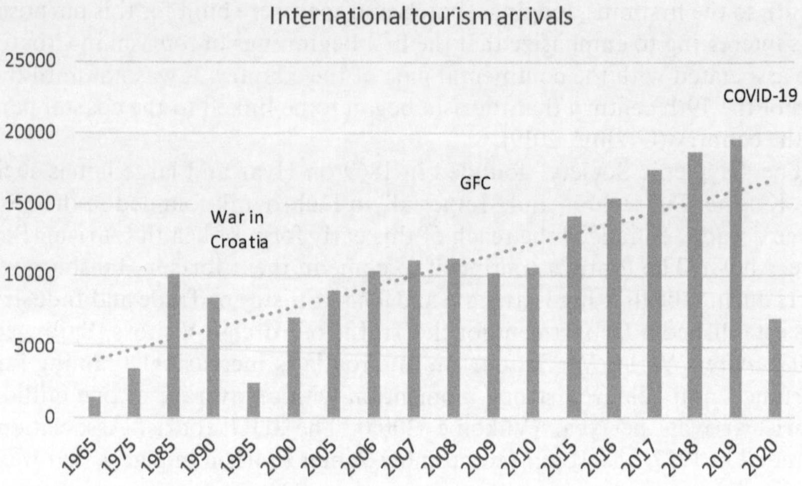

Figure 7.1 International tourism arrivals (in 000) in Croatia.

Table 7.1 Role of tourism in Croatia

	2019		
	Croatia	EU	World
Contribution of travel & tourism to GDP	25% of total economy	9.5% of total economy	10.3% of total economy
Contribution of travel & tourism to employment	25.1% of total employment	11.2% of total employment	1 in 10 jobs around the world
International visitor impact	38.6% of total exports	6.2% of total exports	6.8% of total exports

Source: Adapted from World Travel & Tourism Council (2021). *Economic Impact Reports*. Available at: https://wttc.org/Research/Economic-Impact (Accessed: 5 May 2021).

According to Dwyer et al. (2017) "in terms of tourism, Croatia is a rising star. Although occupying only 1.3% of EU territory, and accounting for less than 1% of the total EU population, it realizes 61 million of tourist nights or 5.1% of EU total. When the number of tourist nights is put in proportion to the population size, Croatia is the most popular destination in the European Union, alongside Malta and Cyprus." The fact that Croatia is a very developed tourist destination but also a country whose economy is strongly dependent on tourism, especially if compared to the global level, is confirmed by the data in Table 7.1.

While globally the contribution of travel and tourism to GDP is equal to 10.3% of total economy and at EU level 9.5% of total economy, in Croatia this contribution reaches a 25% of total economy. A very similar situation exists with contribution of travel and tourism to employment. The strength of tourism in Croatia is even more pronounced if one looks at international visitor impact. At the global level, international visitor impact amounts to 6.8% of total exports, at the EU level of 6.2% of total exports, while at the level of Croatia it amounts to a high 38.6% of total exports.

A meaningful assessment of the tourism sector of a particular country requires a brief overview of the structure of accommodation capacities as well as the capacity utilization of accommodation capacities.

Croatia has a very unfavourable structure of accommodation capacities. At the EU level, the share of beds in hotels and similar accommodation in the observed period is the standard 44%, while at the Croatian level it declines and fell from 35% in 2011 to 15% in 2019.

It is very interesting to compare the capacity utilization in the EU in relation to Croatia.

A comparison of the capacity utilization shows that Croatia lags far behind the rest of the EU. While at the EU level the total capacity utilization in the observed period increased from 13.35% in 2011 to 14.49% in 2019, at the Croatian level it decreased from 2.42% in 2011 to 1.67% in 2019. Analysing capacity utilization of hotels and similar accommodation, growth is visible at both the EU and Croatian levels. At the EU level, it was 18.30%

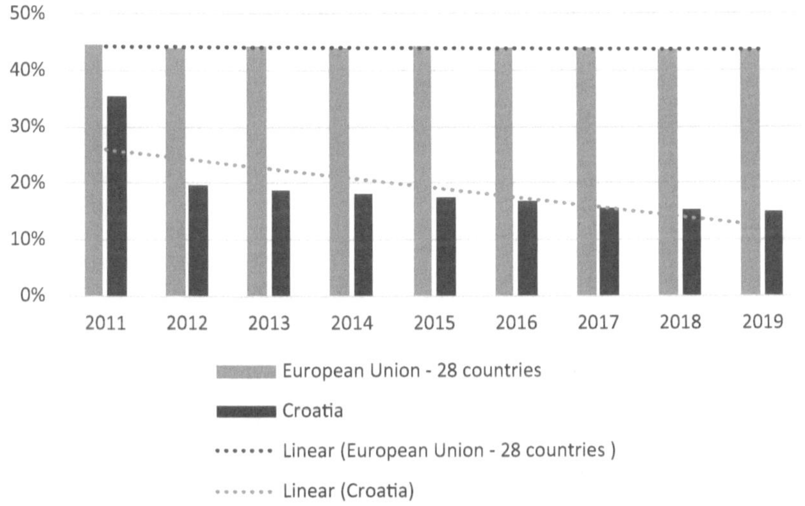

Figure 7.2 Share of bed in hotels and similar accommodation in total accommodation, EU vs. Croatia (2011–2019).

Source: Author's calculation according to EUROSTAT Data Browser (2021a). *Net occupancy rate of bed-places and bedrooms in hotels and similar accommodation (NACE Rev. 2, I, 55.1) by size class*. Available at: https://ec.europa.eu/eurostat/databrowser/view/tour_occ_anor/default/table?lang=en (Accessed on: 15 April 2021); EUROSTAT Data Browser (2021b). *Number of establishments, bedrooms and bed-places*. Available at: https://ec.europa.eu/eurostat/databrowser/view/tour_cap_nat/default/table?lang=en (Accessed on: 15 April 2021).

in 2011, while in 2019 it reached a level of 19.57%. At the level of Croatia, it amounted to 4.27% in 2011, while in 2019 it reached the level of 4.40%. Looking at figures it is quite clear how hotel and similar accommodation achieves higher utilization of accommodation capacities and confirms the existence of an unfavourable structure of accommodation capacities in Croatia. The disadvantage of the structure of accommodation capacities is even more evident from Figure 7.4.

Hotels participate with only 13.3% in the total accommodation capacity. Despite this, hotels account for 28.3% of all overnight stays. Back in 2010 Horwath HTL (2010) pointed out that in Croatia there is a high share of complementary types of accommodation, i.e., capacity with a lower average price than hotels, which has an adverse effect on income. Ten years later, it is possible to

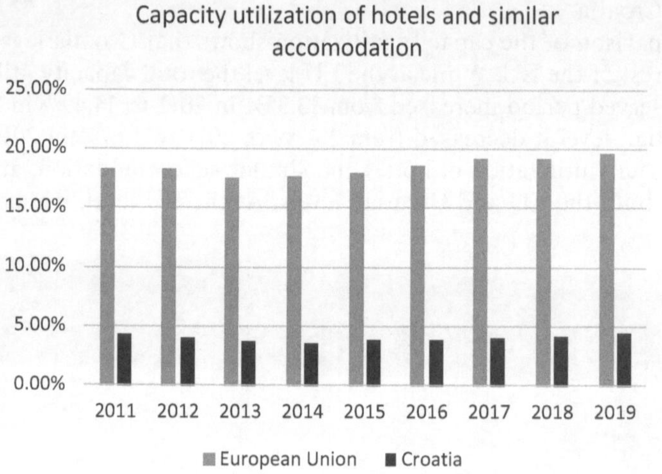

Figure 7.3 Capacity utilization, EU vs. Croatia (2011–2019).
Source: Author's calculation according to EUROSTAT Data Browser (2021c). *Nights spent at tourist accommodation establishments.* Available at: https://ec.europa.eu/eurostat/databrowser/view/tour_occ_ninat/default/table?lang=en (Accessed on: 15 April 2021).

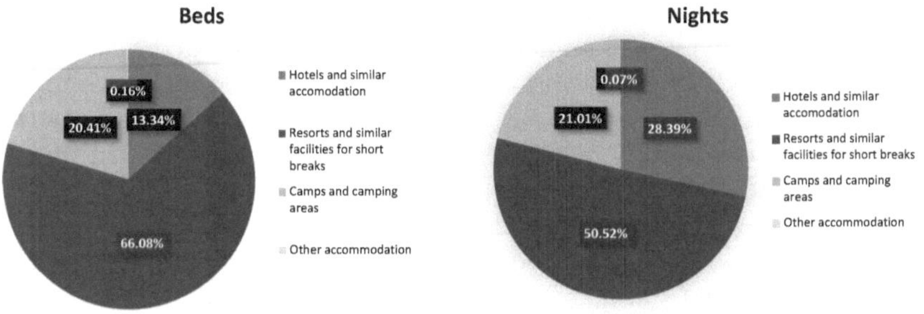

Figure 7.4 Structure of accommodation capacities and structure of overnight stays by types of accommodation facilities in 2019.

Source: Author's calculations according to Bureau of Statistics. Available at: https://www.dzs.hr/Hrv/Covid-19/turizam.html (Accessed: 14 April 2021).

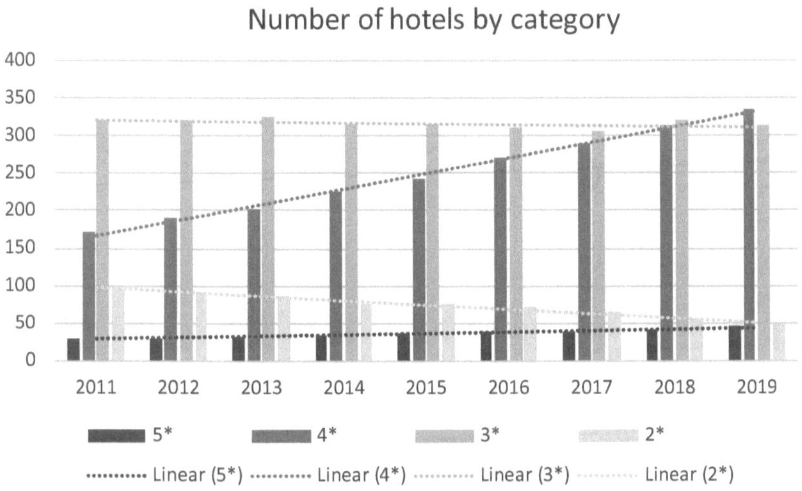

Figure 7.5 Number of hotels by category, Croatia (2011–2019).

Source: Author's calculation according to Ministry of Tourism (2018). *Tourism in numbers 2017*. Available at: https://mint.gov.hr/UserDocsImages/AA_2018_c-dokumenti/180608_HTZ-TUBHR_2017.PDF (Accessed: 14 April 2021); Ministry of Tourism (2020). *Tourism in numbers 2019*. Available at: https://www.htz.hr/sites/default/files/2020-07/HTZ%20TUB%20HR_%202019%20%281%29.pdf (Accessed: 14 April 2021); Ministry of Tourism (2016). *Tourism in numbers 2015*. Available at: https://mint.gov.hr/UserDocsImages/arhiva/TUB2015HR.pdf (Accessed: 14 April 2021); Ministry of Tourism (2014). *Tourism in numbers 2013*. Available at: https://www.htz.hr/sites/default/files/2016-11/Turizam-u-brojkama-2013.pdf (Accessed: 14 April 2021).

conclude that this situation at the Croatian level has worsened because, despite the absolute growth in the number of hotels, the share of accommodation in hotels and similar accommodation has halved compared to 2010.

Additionally, it is interesting to look at the number of hotels according to the categorization in Croatia.

Observing the number of hotels according to the categorization in Croatia, it can be seen that in the observed period the number of 5- and 4-star hotels increased, while the number of 3- and 2-star hotels decreased. This unequivocally indicates an improvement in the quality of hotel accommodation in Croatian tourism. In the observed period, the number of 5-star hotels increased by 58.5%, and the number of 4-star hotels by 95.35%. In the same period, the number of 3-star hotels decreased by 2.18%, while the number of 2-star hotels practically halved and decreased by 51%.

Figure 7.6 shows the accommodation capacities according to the categorization of hotels in Croatia.

The observation of accommodation capacities according to the categorization of hotels shows an identical trend, the number of accommodation capacities increased in 5- and 4-star hotels, and decreased in 3- and 2-star hotels. In hotels with 5-stars, it increased in the observed period by 19.6%, and in 4-star hotels by 54.4%. In 3-star hotels the number of accommodation capacities decreased by 21.8%, and in 2-star hotels by 55.7%.

In the context of every tourist destination, including Croatia, it is important to observe the dynamics of the arrival of guests throughout the year. Given the specific structure of accommodation capacities in Croatia, it is

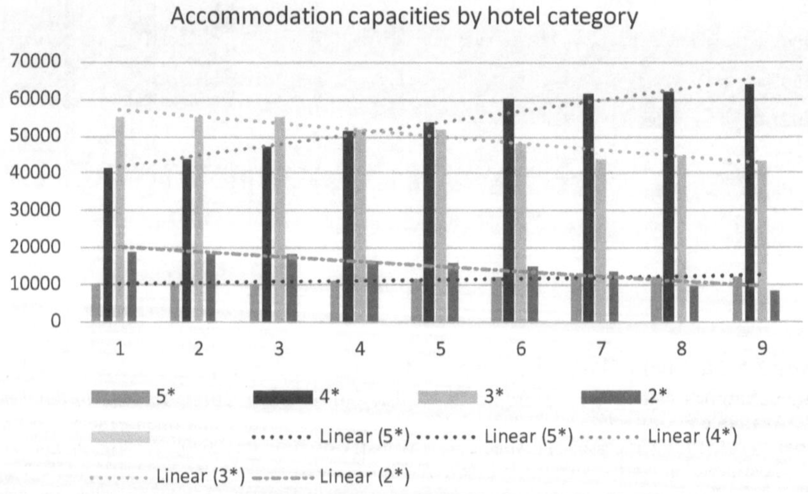

Figure 7.6 Accommodation capacities by hotel category, Croatia (2011–2019).

Source: Author's calculation according to Ministry of Tourism (2018). *Tourism in numbers 2017.* Available at: https://mint.gov.hr/UserDocsImages/AA_2018_c-dokumenti/180608_HTZTUBHR_2017.PDF (Accessed: 14 April 2021); Ministry of Tourism (2020). *Tourism in numbers 2019.* Available at: https://www.htz.hr/sites/default/files/2020-07/HTZ%20TUB%20HR_%202019%20%281%29.pdf (Accessed: 14 April 2021); Ministry of Tourism (2016). *Tourism in numbers 2015.* Available at: https://mint.gov.hr/UserDocsImages/arhiva/TUB2015HR.pdf (Accessed: 14 April 2021); Ministry of Tourism (2014). *Tourism in numbers 2013.* Available at: https://www.htz.hr/sites/default/files/2016-11/Turizam-u-brojkama-2013.pdf (Accessed: 14 April 2021).

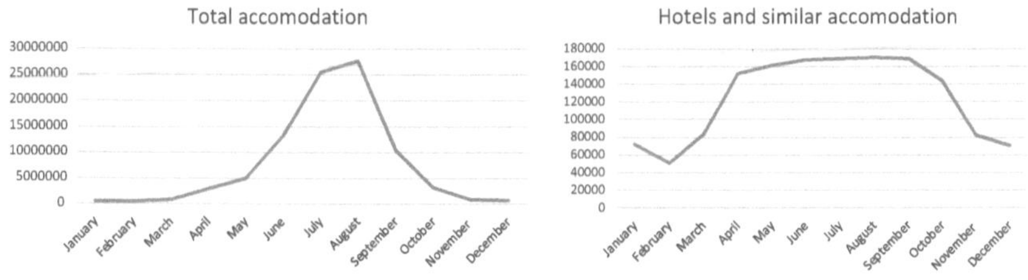

Figure 7.7 Number of tourism arrivals by months, 2019 – total accommodation vs. hotels and similar accommodation.

Source: Adapted from Bureau of Statistics. *Tourism, 2019*. Statistical Reports. Available at: https://www.dzs.hr/Hrv_Eng/publication/2020/SI-1661.pdf (Accessed on 15 April 2021).

very interesting to observe seasonality at the level of total accommodation capacities, but also separately for hotels and similar accommodation capacities, as shown in Figure 7.7.

The presented figures clearly indicate the problems caused by the specific structure of accommodation capacities of Croatian tourism. It is possible to notice, by observing the number of guest arrivals at the level of total accommodation capacities, that Croatian tourism is characterized by high seasonality related exclusively to the summer months. If the number of guest arrivals is observed at the level of hotels and similar accommodation, it is evident that this seasonality is significantly reduced and that this type of accommodation operates at the same level for a full seven months of the year.

Finally, it seems useful to point out the differences in the occupancy of permanent beds that occur when comparing the total accommodation capacity with hotels and similar accommodation.

The occupancy of permanent beds also speaks in favour of hotel accommodation, which throughout the observed year recorded a far higher occupancy compared to the occupancy of total accommodation capacity.

In addition to all of the above, it is not surprising that Croatia's Master Plan and Strategy for the Development of Croatian Tourism until 2020 focused on the scenario of opening to larger development investments, and the strategic priority of attracting investors has been set. Emphasis was placed on the intensified construction of large 4- and 5-star hotel facilities, greater opening to foreign capital, greater dependence on foreign mainstream tour operators and an orientation towards internationally standardized tourist products. Nevertheless, although the number of 4- and 5-star hotels has grown, relatively speaking, the share of hotel accommodation in total accommodation has deteriorated.

Theoretical Considerations – What Are the Benefits from FDI in Tourism?

Tourism is a global phenomenon of today whose further development requires continuous investment of every tourism-oriented country. FDI imposes itself

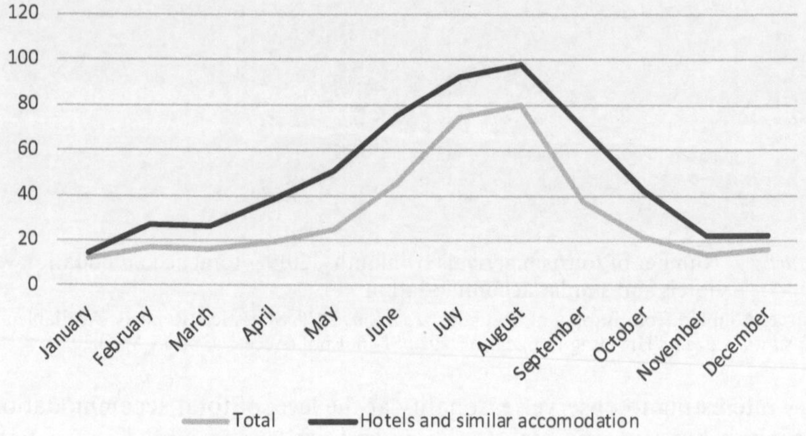

Figure 7.8 Occupancy of permanent beds (%), 2019 – total accommodation vs. hotels and similar accommodation.

Source: Adapted from Bureau of Statistics. *Tourism, 2019*. Statistical Reports. Available at: https://www.dzs.hr/Hrv_Eng/publication/2020/SI-1661.pdf (Accessed on 15 April 2021).

as a significant factor contributing to the further development of tourism, despite the fact that the share of global FDI in tourism is less significant (Endo, 2006; UNCTAD, 2007). FDI creates the preconditions for the integration of host countries into international tourism networks, which in turn can result in increased tourism flows and generate higher revenue from tourism-related activities (Endo, 2006). It should also be emphasized that FDI in tourism largely consists of investment in hotels (UNCTAD, 2007).

A review of previous studies examining the effects of FDI on tourism on the host country suggests that global FDI has a significant impact on tourism itself. As with foreign investments in any other part of the economy, and as far as tourism is concerned, the degree of development of the country also plays a major role in the effects.

Given the fact that Croatia lags behind the amount of hotel accommodation compared to the level of Europe, Table 7.2 shows the most common effects that can lead to the presence of global tourism brands.

One of the most important **quantitative effects** is *the inflow of new capital and the removal of capital constraints.* Further development of tourism in developing and less developed countries often requires long-term assets that are not available in sufficient quantities in these countries. These countries often lack their own long-term capital, and FDI through transnational corporations can help remove capital constraints (Dwyer et al., 2010).

Transnational corporations typically allow for *an increase in employee personal income* because they pay higher wages there are better employers. It should also be emphasized that the more mature the market, the smaller

Table 7.2 The most important effects of global tourism brands

Quantitative economic effects	***Qualitative economic effects***	***Effects on business standards***	***Effects on social and living standards***	***Ecological effects***
New capital inflows and removal of capital constraints **Increasing employment** **Increasing employee personal income** **Increase in tax revenues** **Increasing the number of arrivals of foreign tourists** **Higher utilization of hotel capacities** **Encouraging the arrival of new TNCs in other sectors of the economy**	Transfer of knowledge, skills and technology Foreign exchange outflows Positioning the destination on the global tourist map Raising the image of the destination Greater connectivity with local suppliers than domestic hotels	Introduction of international accounting and financial reporting standards Improving fire and safety standards	Financial support to the improvement of local and social services Influence on cultural identity, transmission of the Western way of life	Introduction of global environmental standards Use of environmentally friendly systems Environmental degradation due to usurpation of large spatial areas

Source: Adapted from Nikšić Radić, M. and Lukinić, M., 2014. Effects of tourism transnational corporations on developing and less developed countries. *Oeconomica Jadertina* 4(2), pp. 53–63.

the pay gap between domestic and local hotels (UNCTAD, 2007). In general, foreign corporations pay 40% more than the average wage than local ones, and the difference is larger in low-income countries in Asia and Latin America (OECD, 2008). Foreign takeovers have led to higher average wages in companies, with very little impact in Germany, 5% in the UK, 8% in Portugal, 11% in Brazil and 19% in Indonesia (OECD, 2008).

Hotels make a significant contribution to *tax revenues*, which can support vital social programs and infrastructure investments that have important effects on the development of individual economies. Hotels operating in tourist-centred economies are likely to have higher direct taxes (IFC, 2016). Hotels generate significant foreign exchange and tax revenues (e.g., payroll tax, VAT, luxury tax, profit tax, property tax, import tax, exit tax, etc.). In some developing countries, the importance of FDI in total tax revenues is very high, which allows the development of measures aimed at reducing poverty (De Schutter et al., 2013).

Of the quantitative effects, it is certainly important to point out the *higher number of international tourism arrivals* in a tourist destination because global hotel brands are connected to the global market and access to global distribution networks (UNCTAD, 2007). Numerous studies to date have shown that higher levels of FDI affect the higher number of international tourism arrivals (Tang et al., 2007; Salleh et al., 2011; Selvanathan et al., 2012; Samimi and Sadeghi, 2013; Perić and Nikšić Radić, 2016a; Bezić and Nikšić Radić, 2017).

Consequently, with a larger number of international tourism arrival, there is *an increase in employment* and income, i.e., the multiplier effects of tourism (Dwyer et al., 2010). It is also important to point out the established pattern of employment by transnational corporations: in the initial years, foreign workers usually hold management positions, and only when the country becomes more experienced in running hotel business can some of the more important positions be taken by domestic workers (Dunning and Kundu, 1982).

Foreign-owned hotel complexes record higher utilization of hotel capacities. In general, the occupancy rate of foreign hotels is 5–20% higher than domestic hotels, but in mature tourist destinations it may be lower (UNCTAD, 2007). Their strength is particularly important during periods of economic recession (O'Neill and Carlbäck, 2011).

Branded luxury hotels are important because in addition to raising the image of the country, they promote the destinations in which they are located which *automatically increases interest in foreign investment in other sectors of the economy*, and encourages the arrival of FDI in other sectors. Given the fact how sensitive tourism is, the arrival of hotel brands is a kind of safety signal to other investors as well.

The qualitative **effects** of hotel brands are also visible through the transfer of so-called best practices, and it is possible to conclude that there is a *transfer of technology, knowledge and skills*. One of the main arguments for (and benefits from) FDI is technology transfer including managerial expertise. Foreign companies can make a "demonstration effect" on domestic entrepreneurs. In developing destinations, an increased level of management skills has been crucial to meet the demands of foreign tourists and to maintain the international competitiveness of the local product (ESCAP, 1994). Training programs conducted by foreign-owned companies may be treated as a form of "technology transfer," creating the benefits of spillovers to domestic companies due to a more skilled workforce within the tourism sector.

The negative effect that is very often attributed to FDI is the occurrence of *foreign exchange outflows*. The issue of outflows must be analysed in the context of the potential that tourism development leads to, which is the generation of higher income and employment in the host country (UNCTAD, 2007). Managerial fees and profits associated with FDI are leaving the host country by reducing the economic contribution of tourism expenditures, i.e., there is a repatriation of profits. This is often treated as a "outflow" from the host country due to FDI whereby the net use for the country from tourism is reduced by an appropriate amount. Nevertheless, such a view rests on a misconception because these profits are not "lost" for the host country in any significant sense, given that the profit would not even exist if the investor did not first bring in capital to invest (Dwyer et al., 2010).

Qualitative effects of transnational corporations in tourism are certainly *positioning the destination on the global tourism map* and *raising the image of the destination* because the involvement of corporations in marketing and promotional activities can increase awareness of the destination (UNCTAD, 2007). The role of well-known marketing brands with a global marketing reach can be particularly important for less developed countries that have limited resources or opportunities to promote their destinations. Well-known hotel brands of a certain quality may provide additional security for potential tourists.

An important effect of FDI in hotels is also the fact that foreign hotels are characterized by a *larger connection with local suppliers than domestic hotels.* Global hotel brands sometimes make greater efforts than domestic hotels to connect with local suppliers (UNCTAD, 2007). Such hotels conduct ongoing procurement locally, especially of food products, whenever there is an adequate and secure supply of the correct quality at a competitive price (UNCTC, 1992).

Effects on business standards include the application of business rules that apply in the international market and include the introduction of international accounting and financial reporting standards as well as the improvement of fire and safety standards. TNCs use more advanced management and financial systems, and in this field there is a significant difference between TNCs and domestic hotels even in mature destinations (UNCTAD, 2007).

TNCs can affect both **social and living standards**. Global hotel chains set high standards of social responsibility by donating goods and services to the local community (Abdo and Mann, 2010) and are involved in successful best practice community-linked businesses, driven by social service (Davidson and Sahli, 2015). They are also often credited with influencing cultural identity as an inevitable consequence of globalization.

As a final point, it is certainly necessary to emphasize the importance of **environmental effects**, especially due to the fact Croatia is a full member of the EU. Buildings (including hotels) account for 40% of total energy consumption in the EU (Directive 2010/31/EU). The EU has adopted a climate and energy package that sets ambitious energy and climate targets for a 20% reduction in CO_2 emissions, a 20% reduction in energy consumption and a 20% increase in renewable energy by 2020. This will require the installation of expensive renewable plants in hotels as well. Many hotels in the EU have already significantly reduced energy consumption by becoming partially or completely independent of conventional energy resources (Hotel Energy Solutions, 2011). The same procedure awaits hotels in Croatia, and the entry of international brands would be a guarantee of sustainable development, because such investors have the financial resources and experience in introducing alternative energy sources. They are also more aware of new global environmental standards, introduce environmentally friendly systems and have a high environmental ethic (UNCTAD, 2007; Davidson and Sahli, 2015).

The negativity that accompanies global hotel chains is the need for large spatial areas, which leads to *environmental degradation*. However, the destruction of the valuable natural environment and the alienation of the local population is contributed by large-scale enclave-type facilities, both foreign-owned and domestically owned. In Croatia, the uncontrolled development of private accommodation also contributes to the destruction of the environment. The above leads to the conclusion that the problem lies not so much in the origin of property but more in the nature of planning and urban laws that allow such construction. TNCs may be even more aware of the need to build and operate content that will comply with international and local environmental standards than domestically-owned enterprises (UNCTAD, 2007).

FDI in Croatian Economy and Croatian Tourism

Geographically, historically and culturally, Croatia is an integral part of Europe, is an official member of the EU and in its development seeks to approach the level of development of developed Western countries.

The inflow of FDI in the period from 1993 to 2011 was influenced by various factors: the Homeland War in Croatia, entry into the World Trade Organization, involvement in European integration processes such as the Stability Pact, the Stabilization and Association Agreement with the EU, CEFTA membership and the opening of EU membership negotiations. Involvement in European integration processes was also reflected in an increase in FDI inflows. Then in 2009, when the global financial crisis hit Croatia, there was a drastic drop in FDI. In the observed period, until 2008, a rapid increase

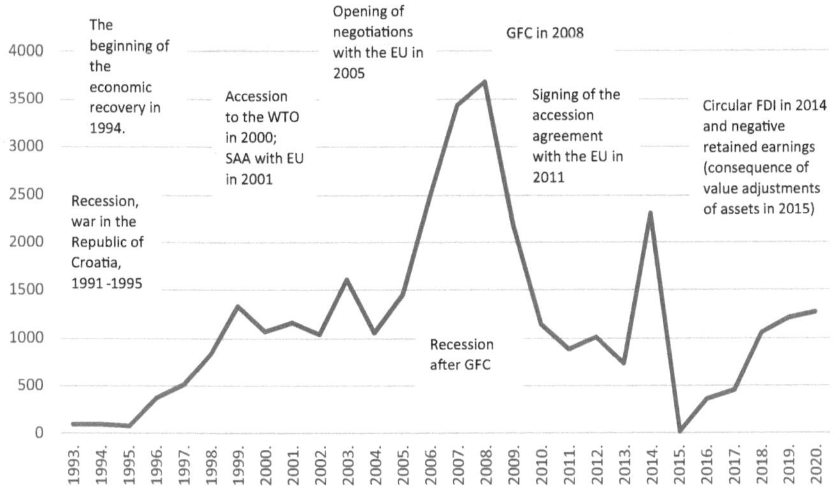

Figure 7.9 Total FDI in Croatia.

Source: Author's calculation according to Croatian National Bank Data. Available at: https://www.hnb.hr/statistika/statisticki-podaci/sektor-inozemstva/inozemna-izravna-ulaganja (Accessed: 4 April 2021). and Kersan-Škabić, I. and Zubin, C., 2009. The influence of foreign direct investment on the growth of GDP, on employment and on export in Croatia. *Ekonomski pregled, 60*(3–4), pp.119–151.

in FDI inflows was recorded, but one of the reasons for this is the financing of the state budget deficit. The state sold its ownership stakes in individual companies, but did not use the money to encourage new investments but to save the budget due to the problem with deficit. The signing of accession negotiations with the EU in 2011 again boosted FDI growth, albeit in a minor way. Croatia ended 2013 with the lowest FDI inflow since 1995. The weak inflow of foreign capital in the period from 2008 to 2013 determines a prolonged period of stagnation of the Croatian economy in an environment of uncertainty in the global financial markets. This is a consequence of the globally high aversion to risk, unfavourable investment environment, but also the absence of privatization processes that greatly contributed to the inflow of foreign capital in pre-crisis times. In 2015, Croatia fell to war levels of FDI inflows because of circular FDI in 2014 and negative retained earnings (consequence of value adjustments of assets in 2015). After that, in 2015, Croatia again recorded the growth of FDI until 2020, when the entire world market was hit by the COVID-19 situation, and UNCTAD (2020) expects that the decline in FDI at the global level will be some 40%.

Looking at FDI by activity orientation, the largest amount of investments in Croatia in the period from 1993 to 2020 was directed to financial intermediation (except for insurance and pension funds). The orientation of investments by activities is shown in Table 7.3.

Table 7.3 FDI by activities (in millions of euros), 1993–2020

Activities	*Total*	*% in total*
Financial service activities, except insurance and pension funding	7345.276	22.23
Wholesale trade, except of motor vehicles and motorcycles	2845.832	8.61
Equity investments in real estate	2422.12	7.33
Real estate business	2162.017	6.54
Retail trade, except of motor vehicles and motorcycles	2037.78	6.17
Telecommunications	2017.288	6.10
Accommodation	1501.187	4.54
Manufacture of coke and refined petroleum products	1265.997	3.83
Construction of buildings	1177.803	3.56
Manufacture of basic pharmaceutical products and pharmaceutical preparations	1115.822	3.38
Other activities	8626.313	27.70
Total	33044.78	100.00

Source: Author's calculation according to Croatian National Bank Data. Available at: https://www.hnb.hr/statistika/statisticki-podaci/sektor-inozemstva/inozemna-izravna-ulaganja (Accessed: 4 April 2021).

Most investments were made in financial intermediation, except for insurance and pension funds (EUR 7.3 billion or 22.2%), wholesale trade and trade intermediation (EUR 2.8 billion or 8%, 6%), real estate (EUR 2.4 billion or 6.5%), retail trade, except motor vehicles and motorcycles (EUR 2.0 billion or 6.2%), telecommunications (EUR 2.0 billion or 6.1%), **accommodation (EUR 1.5 billion or 4.5%)**, production of coke and refined petroleum products (EUR 1.3 billion or 3.8%), construction of buildings (EUR 1.2 billion or 3, 6%), manufacture of basic pharmaceutical products and pharmaceutical preparations (EUR 1.1 billion or 3.4%) and other activities (EUR 8.6 billion or 27.7%); 72.3% of total investments are accounted for by the ten activities listed.

Figure 7.10 shows the inflow of FDI into Croatian tourism, which in fact refers to the inflow of foreign capital into the tourism and hospitality industry.

Most of the FDI in Croatian tourism in the first half of the 1990s was related to the takeover and/or creation of ownership portfolios from existing companies, i.e., to the speculative acquisition of large land areas in the coastal zone in the period from the end of war operations to 2000. FDIs were mainly concentrated on the reconstruction of existing capacities because they are mainly located in the most attractive and acceptable urban locations and their reconstruction seeks to maximize the location advantage (Družić, 2010). Looking at FDI in tourism in the period from 2001 to 2012, it is possible to conclude that such investments were very sporadic in the observed period, and in particular took on negative characteristics after 2008 with the onset of the economic crisis. The average growth rate of FDI in tourism in the period from 2001 to 2011 was 2.3%.

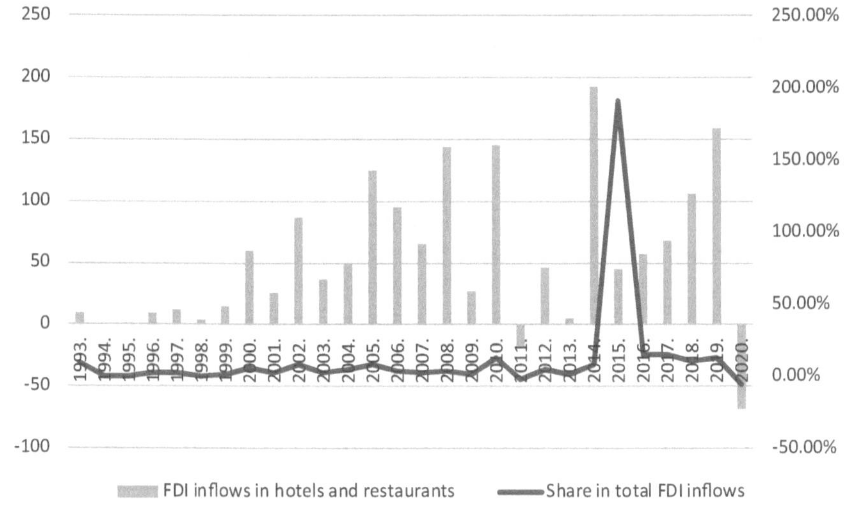

Figure 7.10 FDI in tourism of Croatia.

Source: Author's calculation according to Croatian National Bank Data. Available at: https://www.hnb.hr/statistika/statisticki-podaci/sektor-inozemstva/inozemna-izravna-ulaganja (Accessed: 4 April 2021).

At the end of 2012, the Master Plan and Strategy for the Development of Croatian Tourism was adopted, which set the vision that by 2020 Croatian tourism would be recognized and acknowledged as a priority economic activity (Government of the Republic of Croatia (2013).). In the mentioned study, one of the development scenarios was set, and that is the opening to larger development investments, which put more emphasis on the construction of new accommodation and other luxury capacities of the tourist offer, especially in larger cities on the coast, but also in larger complexes outside existing urban units. Thus, the need for greater opening of Croatia to foreign investors and increased construction of large new hotels and other major tourist accommodation facilities was clearly emphasized. With the adoption of these decisions, FDI in Croatian tourism in relation to total FDI increased from 3.6% (period from 1993 to 2012) to 7.6% (period from 2012 to 2020).

The numbers are more than clear and government initiatives aimed at attracting FDI to tourism have been identified as important for its further tourism development (Hospitality, 2012). However, if one looks at how many renowned hotel brands Croatian tourism can boast of today, it is clear that it lags far behind the Mediterranean competition. The presence of global hotel brands is very modest. For example, Croatia still cannot boast of the presence of a single Hyatt. It is possible to conclude that the efforts made so far to attract renowned hotel brands have resulted in limited results.

Achieved Effects of FDI in Croatian Tourism

The effects of previous FDIs in tourism can be summarized as a result of mapping the effects from the micro level to the macro level. Research on the economic effects of FDI in Croatian tourism, following the example of the International Financial Association, starts from the micro level (Abdo and Mann, 2010; ODI, 2012). Perić and Nikšić Radić (2016) chose as a research object, the FDI in a hotel located in the area of small local government, in the continental part of Croatia. Finding a foreign-owned hotel located in a smaller municipality in which there is no presence of other hotels has enabled a more precise identification of economic effects and more reliable drawing of research conclusions. The analysis of the selected hotel showed that a foreign-owned hotel, in a less developed country such as Croatia, has a significant development effect on the local economy, which is in line with previous research (Abdo and Mann, 2010; ODI, 2012). A summary of the recognized effects of FDI on tourism at the micro level is shown in Figure 7.11.

Although in the observed case a loss of domestic ownership occurred, the entrance of a foreign investor brought many positive effects. The small municipality developed in the midst of the financial crisis. The new owner

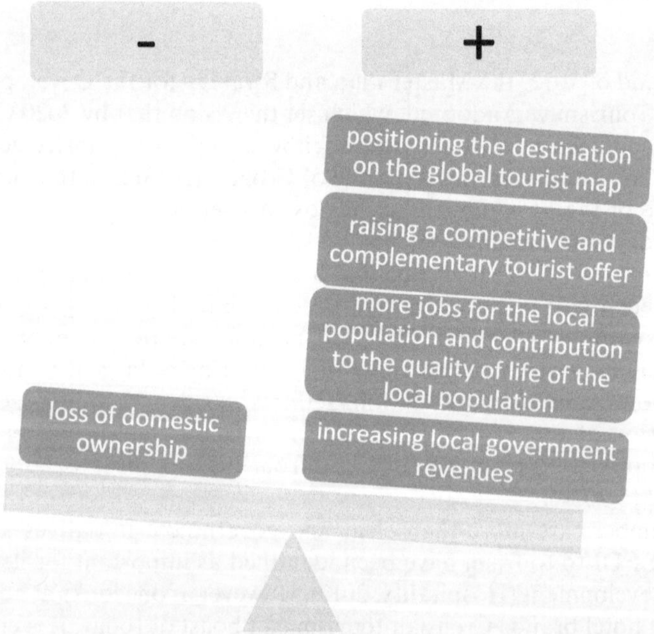

Figure 7.11 Impacts of FDI in tourism on micro level.
Source: Adapted from Perić, J. and Nikšić Radić, M., 2016b. Development Impact of FDI in Hotel: Case Study of Terme Tuhelj in Croatia. *Almatourism – Journal of Tourism, Culture and Territorial Development,* 7(13), pp. 66–78. doi:10.6092/issn.2036-5195/5983.

initiated the development of tourism, the development of competitive and complementary activities, preserved existing jobs, created new jobs and the investor contributed in many ways to the local environment through numerous sponsorships and donations. In addition to the direct effects of this investment, there are also indirect effects in the form of consumption of guests outside the hotel and the effects caused in the form of consumption of hotel employees' salaries.

Furthermore, it is logical to assume that there is a mapping of effects from the micro level to the macro level as shown by the following scheme.

These effects of FDI on Croatian tourism have been proven empirically. Perić and Nikšić Radić (2015) proved that FDI in tourism has a positive significant effect on tourism productivity, i.e., that FDI in tourism influenced the growth of tourism productivity in the observed period. Furthermore,

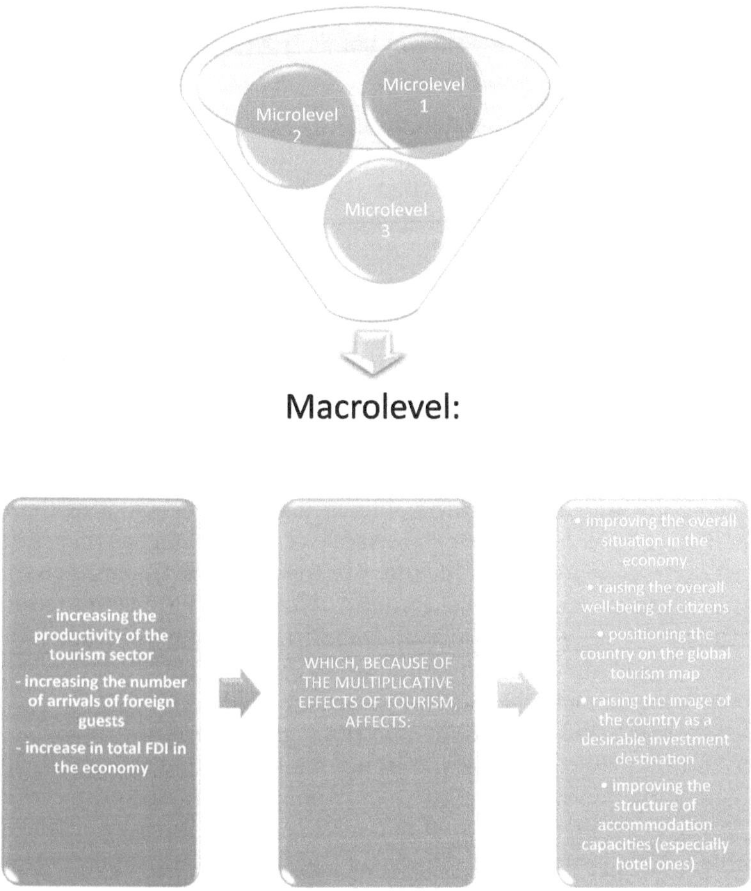

Figure 7.12 Impact of FDI in tourism on Croatia tourism – Mapping effects from micro level to macro level.

Bezić and Nikšić Radić (2017) proved the causal link between FDI in tourism and the gross value added of Croatian tourism both in the long run and in the short run. It has also been proven that FDI in tourism influenced the number of arrivals of foreign guests in Croatia (Perić and Nikšić Radić, 2016a). FDI in tourism also affected the overall inflow of FDI into the Croatian economy (Nikšić Radić, 2014).

Given that tourism in general plays a strong multiplicative role in the economy, it is possible to conclude that FDI in Croatian tourism so far improved the overall situation in the economy. FDI raised the overall well-being of citizens, positioned the country on the global tourism map, raised the image of the country as a desirable investment destination and improved the structure of accommodation capacities (especially hotel ones). However, this is still far from a satisfactory structure of accommodation capacities and potential that Croatia has.

Croatia Still Needs FDI in Tourism but Is It Attractive to Foreign Investors?

It is quite clear that the further development of Croatian tourism still needs FDI, especially in terms of attracting global hotel brands.

The main competitors of Croatian tourism in the Mediterranean region are attracting more investment in the hospitality sector than Croatia (Patricolo, 2019). The fact is that Croatia as a tourist destination is becoming increasingly attractive. In 2018, Croatia appeared on Virtuoso's 2018 Luxe Report list of the top emerging destinations: Iceland, Croatia, Japan, Cuba and Portugal (Virtuoso, 2017). In 2021, Virtuoso included Croatia on the list of the world's most desirable destinations alongside South Africa, Italy, Australia, France, Japan, Great Britain, Greece, Argentina, Kenya, Botswana, Spain, Iceland, Portugal, Mexico, New Zealand, Ireland, Tanzania and Zanzibar, Thailand and California (Virtuoso Wanderlist). As a result, "high-end hotel brands" are showing increasing interest in entering the Croatian market (Patricolo, 2019). In the last few years, Croatia has thus become extremely attractive to foreign investors in tourism. The first edition of the Global Tourism Locations of the Future 2017/2018, fDi Magazine (Mullan, 2017), ranked Croatia as the seventh-largest destination with the greatest investment opportunities in tourism. Two year later, Croatia is ranked as the seventeenth-largest destination.

There are many investment opportunities in the Croatian tourism market, still, given that the hotels' supply does not match demand (Doggrell, 2018). Nonetheless, the following should also be pointed out. According to Filip Vucagic, director of Colliers Croatia, Croatia needs to attract greenfield investments, a huge potential for real estate investments in Croatia, especially in tourism because "after the realisation of all brownfield investments – which are the prime choice of investors and still available on the market – we will not have a lot of interest of investors in greenfield projects" (Patricolo, 2019).

What to do about it? In 2016, Perić and Nikšić Radić (2016c) recognized that political stability, stimulating macroeconomic business conditions (including ranking at Doing Buissnes), removal of administrative and legislative barriers, removal of the country's image as a corrupt destination and educating tourism staff at all levels are of particular importance for FDI in Croatian tourism. All of the above were the main obstacles to the entry of foreign capital into the Croatian tourist market. High seasonality that dominates the Croatian tourism market as well as poor positioning on the Doing Business list (58th overall and 159th in dealing with construction permits) are still the main barriers to entry of foreign investors (Horwath, 2019) and the efforts of government officials in Croatia must be continuously focused on improving such long-standing domestic issues.

Finally, it seems convenient to quote Fox (2014) and her view on Croatia: "Invest in an unexplored destination and maintain it as a luxury property. The investment is a good one whether you decide to sell or keep."

Conclusion

The analysis of Croatian tourism in this chapter unequivocally points to the poor structure of accommodation capacities in relation to other EU member states and the extremely low representation of hotel accommodation. In addition, Croatia is characterized by a very low presence of global hotel brands in relation to other competitive tourist destinations, i.e., insufficient attraction of FDI in tourism.

Even at the time of ex-Yugoslavia, there was a downward trend in overnight stays by foreign tourists in hotels and a growing trend for overnight stays in other types of tourist accommodation. "In 1965 51.1% of foreign visitors stayed in hotels; ten years later only 47.3%, and, in the 1980s, only 41.9% of foreign tourists stayed in hotels" (Gosar, 1989).

In addition to such a legacy of the structure of accommodation capacities, a problem arose during the development of Croatian tourism due to the quality of hotel accommodation. During the Homeland War, many hotel complexes on the Adriatic coast were abandoned, damaged and many of them were owned by the state. Croatia, along with Slovenia and Belarus, is the country with the highest number of state-owned enterprises in Central and South-Eastern Europe. The number of state-owned enterprises is about 260 enterprises per 1 million inhabitants (Richmond et al., 2019). State-owned enterprises are not only present in natural monopolies, but are also present in manufacturing for retail markets, construction and hotels (EBRD, 2020). Poorly implemented privatization of such state-owned tourist facilities and deindustrialization have led to a decline in competitiveness and loss of achieved spatial planning standards (Kranjčević and Hajdinjak,

2019), which further affected the uncontrolled apartmentization of the Croatian coast and poor accommodation structure.

The adoption of the Development Strategy of Croatian Tourism by 2020 officially determined the Croatian commitment to attracting FDI in tourism, with special emphasis on global hotel brands. With such a political orientation, Croatia has so far managed to attract a significant number of hotel brands but still, not enough to be in line with the EU in the structure of accommodation capacities. All previous research on the effects of FDI on Croatian tourism only further emphasizes the need for such a commitment. FDI in Croatian tourism affected its productivity, the number of international tourism arrival and the overall level of FDI in the Croatian economy.

Today, Croatia is an established tourist destination with an over-tourism burden on its back. This is a consequence of the poor structure of accommodation capacities, which makes Croatian tourism extremely seasonal and thus under the pressure of over-tourism in the summer months. Further, attracting FDI to tourism, especially global hotel brands, Croatia could correct such a situation because it is luxury hotel accommodation that has far less seasonality. It is a country that is very attractive both as a tourist and as an investment destination, and in order to realize this potential, it is necessary to urgently solve its long-standing problem – poor positioning on the Doing Business list.

References

Abdo, H. and Mann, S. (2010). *Development Impacts in the Tourism Sector through the Lens of Two IFC Hotel Investments: What They Show and What We Can Learn - Hyatt Regency in Kiev & Peru Orient Express Hotels*. Washington: IFC, World Bank Group. Available at: http://www.ifc.org/wps/wcm/connect/d26122004c5e66d3837bc7f81ee631cc/Tourism+Sector+Development+Impact+Case+Studies+2010.pdf?MOD=AJPERES (Accessed on: 21 April 2014)

Allcock, J. B. (1991). Yugoslavia. In *Tourism and Economic Development in Eastern Europe and the Soviet Union*. Derek Hall (ed.), pp. 236–258. London: Belhaven Press.

Ballinger, P. (2003). Selling Croatia or selling out Croatia? Tourism, Privatization and Coastal Development Issues in a "New" Democracy. Washington: The National Council for Euroasian and East European Research.

Bezić, H. and Nikšić Radić, M. (2017). Tourism Foreign Direct Investment Led Tourism Gross Value Added: A Co-Integration and Causality Analysis of Croatian Tourism. *Ekonomska istraživanja*, *30*(1), pp. 1443–1460, doi:10.1080/1331677X.2017.1340173.

Bureau of Statistics Database (2021) Arrivals and overnight stays of domestic and foreign tourists, Croatia, monthly. Zagreb Available at: https://www.dzs.hr/PXWeb/Table.aspx?layout=tableViewLayout1&px_tableid=BS_T11.px&px_path=Turizam__Dolasci%20i%20no%c4%87enja%20turista%20u%20komerecijalnim%20smje%c5%a1tajnim%20objektima&px_language=hr&px_db=Turizam&rxid=fb39b06b-9de5-4157-8ae3-1e9f68ef0c80 (Accessed: 14 April 2021).

Bureau of Statistics. Available at: https://www.dzs.hr/Hrv/Covid-19/turizam.html (Accessed: 14 April 2021).

Bureau of Statistics. *Tourism, 2019*. Statistical Reports. Available at: https://www.dzs.hr/Hrv_Eng/publication/2020/SI-1661.pdf (Accessed on 15 April 2021).

Cattaruzza, A. and Sintès, P. (2012). A*tlas géopolitique des Balkans. Un autre visage de l'Europe*. Paris: Autrement.

Croatian National Bank Data. Available at: https://www.hnb.hr/statistika/statisticki-podaci/sektor-inozemstva/inozemna-izravna-ulaganja (Accessed: 4 April 2021).

Davidson, L. and Sahli, M. (2015). Foreign Direct Investment in Tourism, Poverty Alleviation, and Sustainable Development: A Review of the Gambian Hotel Sector. *Journal of Sustainable Tourism*, *23*(2), pp. 167–187, doi:10.1080/09669582.2014.957210.

De Schutter, O., Swinnen, J. and Wouters, J. (2013). *Foreign Direct Investment and Human Development: The Law and Economics of International Investment Agreements*. Oxon: Routledge.

Doggrell, K. (2018). Strong demand drives global investment in Croatia. Available at: https://www.hotelmanagement.net/own/strong-demand-drives-global-investment-croatia (Accessed: 20 May 2021).

Družić, I. (2010). Dugoročni trendovi i ograničenja hrvatskog turističkog tržišta. *Ekonomski pregled*, *61*(3–4), pp. 103–136.

Dunning, J. and Kundu, S. K. (1982). *Impact of Transnational Hotel Chains and Management Companies Operating in Developing Countries in Transnational Corporations in International Tourism*, New York and Geneva: United Nations. Available at: http://unctc.unctad.org/data/e82iia9b.pdf (Accessed on: 20 April 2021).

Dwyer, L., Tomljenović, R. and Čorak, S. (2017). *Evolution of Destination Planning and Strategy – The Rise of Tourism in Croatia*. New York: Palgrave Macmillan.

Dwyer, L., Forsyth, P. and Dwyer, W. (2010). *Tourism Economics and Policy*, Cheltenham: Channel View Publications.

EBRD (2020). Turning Around State-Owned Enterprises in Croatia. *Law in Transition Journal*. Available at: https://www.ebrd.com/publications/law-in-transition-2020-turning-around-state-owned-enterprises.pdf (Accessed: 15 May 2021).

Endo, K. (2006). Foreign direct investment in tourism – flows and volumes. *Tourism Management*, *27*(4), pp.600–614.

ESCAP (1994). *Transnational Corporations and Technology Transfer in Export Processing Zones and Science Parks*. New York: UN.

European Commission (2010). Directive 2010/31/EU of the European Parliament and of the Council of 19 May 2010 on the energy performance of buildings (recast). *Official Journal of the European Communities, Chapter 12 Volume 003*, p. 124–146.

EUROSTAT Data Browser (2021a). Net occupancy rate of bed-places and bedrooms in hotels and similar accommodation (NACE Rev. 2, I, 55.1) by size class. Available at: https://ec.europa.eu/eurostat/databrowser/view/tour_occ_anor/default/table?lang=en (Accessed on: 15 April 2021).

EUROSTAT Data Browser (2021b). Number of establishments, bedrooms and bed-places. Available at: https://ec.europa.eu/eurostat/databrowser/view/tour_cap_nat/default/table?lang=en (Accessed on: 15 April 2021).

EUROSTAT Data Browser (2021c). Nights spent at tourist accommodation establishments. Available at: https://ec.europa.eu/eurostat/databrowser/view/tour_occ_ninat/default/table?lang=en (Accessed on: 15 April 2021).

Fox, J. (2014). Luxury hotels in Croatia. Available at: https://www.luxurytraveladvisor.com/destinations/luxury-hotels-croatia (Accessed: 20 May 2021).

Gosar, A. (1989). Structural Impact of International Tourism in Yugoslavia. *GeoJournal, 19*(3), pp. 277–283.

Government of the Republic of Croatia (2013). Tourism development strategy of the Republic of Croatia until 2020. Available at: https://mint.gov.hr/UserDocsImages/dokumenti/Strategija-turizam-2020-veljaca2013.pdf (Accessed: 20 May, 2021).

Grandits, H. and Taylor, K. (eds.) (2010). *Yugoslavia's Sunny Side: A History of Tourism and Socialism (1950s–1980s)*. Budapest: CEU Press.

Gržinić, J. (2019). *Uvod u turizam - povijest, razvoj, perspective*. Sveučilište Jurja Dobrile u Puli. Pula

Horwath HTL (2010). Analysis of the hotel business in Croatia with proposals for measures to improve the state of the sector. Udruga poslodavaca u hotelijerstvu Hrvatske; Hrvatska gospodarska komora; Zagreb.

Horwath HTL (2019). European chains & hotels report 2019. Available at: https://corporate.cms-horwathhtl.com/wp-content/uploads/sites/2/2019/03/HTL_2019_EU_CHAINS-2.pdf (Accessed 20 April 2021).

Hospitality (2012). The good, the bad and the opportunity in Croatia. Available at: https://hospitality-on.com/en/investment/good-bad-and-opportunity-croatia (Accessed: 2 April 2021).

IFC (2016). A hotel is not just a place to sleep. Available at: https://www.ifc.org/wps/wcm/connect/2e1ff2b5-4863-4bf6-b84b-e4dfd88df4c7/Tourism_FINAL_web.pdf?MOD=AJPERES&CVID=lfJk9wF (Accessed: 18 May 2021)

Kersan-Škabić, I. and Zubin, C. (2009). The influence of foreign direct investment on the growth of GDP, on employment and on export in Croatia. *Ekonomski pregled, 60*(3–4), pp. 119–151.

Kranjčević, J. and Hajdinjak, S. (2019). Tourism Urbanization in Croatia. The Cases of Poreč in Istria and Makarska in Dalmatia. *Südost-europa Journal of Politics and Society, 67*(3), pp. 393–420, doi:10.1515/soeu-2019-002.

Ministry of Tourism of the Republic of Croatia (2013). Tourism development strategy of the Republic of Croatia until 2020. Available at: http://iztzg.hr/UserFiles/Pdf/Izvjestaj-01-Strategija-razvoja-turizma-RH.pdf (Accessed: 20 February 2017).

Ministry of Tourism (2014). Tourism in numbers 2013. Available at: https://www.htz.hr/sites/default/files/2016-11/Turizam-u-brojkama-2013.pdf (Accessed: 14 April 2021).

Ministry of Tourism (2016). Tourism in numbers 2015. Available at: https://mint.gov.hr/UserDocsImages/arhiva/TUB2015HR.pdf (Accessed: 14 April 2021).

Ministry of Tourism (2018). Tourism in numbers 2017. Available at: https://mint.gov.hr/UserDocsImages/AA_2018_c-dokumenti/180608_HTZTUBHR_2017.PDF (Accessed: 14 April 2021).

Ministry of Tourism (2020). Tourism in numbers 2019. Available at: https://www.htz.hr/sites/default/files/2020-07/HTZ%20TUB%20HR_%202019%20%281%29.pdf (Accessed: 14 April 2021)

Mullan, C. (2017). fDi tourism locations of the future 2017/18 – editor's choice awards. Available at: https://www.fdiintelligence.com/Locations/Americas/Colombia/fDi-Tourism-Locations-of-the-Future-2017-18-Editor-s-Choice-Awards (Accessed: 2 March 2021).

Nikšić Radić, M., 2014. The economic impact of foreign direct investment on tourism in Croatia (Doctoral thesis). Rijeka: University of Rijeka.

Nikšić Radić, M. and Lukinić, M., 2014. Effects of Tourism Transnational Corporations on Developing and Less Developed Countries. *Oeconomica Jadertina*, *4*(2), pp. 53–63.

ODI (2012). Investing in hotels and demonstrating development impact: case study of IFC's investment in the Serena Kigali hotel. Available at: http://www.ifc.org/wps/wcm/connect/2e8ac3004b885faebb8ffbbbd578891b/Serena+Kigali+CS.pdf?MOD=AJPERES (Accessed: 20 June 20 2018)

OECD (2008). *Do Multinationals Promote Better Pay and Working Conditions?* Paris: OECD Employment Outlook.

O'Neill, J. and Carlbäck, M. (2011). Do Brands Matter? A Comparison of Branded and Independent Hotels' Performance during a Full Economic Cycle. *International Journal of Hospitality Management, 30*(3), pp. 515–521. doi:10.1016/j.ijhm.2010.08.003.

Orsini, K. and Ostojić, V. (2018). Croatia's tourism industry: beyond the sun and sea. European economy economic briefs. Available at: https://ec.europa.eu/info/sites/default/files/economy-finance/eb036_en.pdf (Accessed: 15 February 2021).

Patricolo, C. (2019). Croatia must promote tourism to attract investment from Mediterranean rivals. Available at: https://emerging-europe.com/business/croatia-must-promote-tourism-to-attract-investment-from-mediterranean-rivals/ (Accessed: 6 April 2021).

Perić, J. and Nikšić Radić, M. (2015). Tourism Productivity and Tourism FDI in Croatia. *Mediterranean Journal of Social Sciences*, *6*(5), pp. 425–433, doi: 10.5901/mjss.2015.v6n5s2p425.

Perić, J. and Nikšić Radić, M. (2016a). FDI-Led Tourism Growth Hypothesis: Empirical Evidence from Croatian Tourism. *European Journal of Tourism, Hospitality and Recreation*, *7*(2), pp. 168–175 doi:10.1515/ejthr-2016-0019.

Perić, J. and Nikšić Radić, M. (2016b). Development Impact of FDI in Hotel: Case Study of Terme Tuhelj in Croatia. *Almatourism – Journal of Tourism, Culture and Territorial Development*, *7*(13), pp. 66–78. doi:10.6092/issn.2036-5195/5983.

Perić, J. and Nikšić Radić, M. (2016c). Macroeconomic Environment and Greenfield Foreign Direct Investment of Hotel Brands. *Ekonomski vjesnik*, *29*(2), pp. 468–485.

Pirjevec, B. (1998). *Ekonomska obilježja turizma*, Zagreb: Golden Marketing.

Richmond, C. et al. (2019). Reassessing the Role of State-Owned Enterprises in Central, Eastern, and Southeastern Europe. *Departmental Papers/Policy Papers, 19*(11), p.1. doi:10.5089/9781498315142.087.

Salleh, N. H. M., R. Othman, and Sarmidi, T. (2011). An Analysis of the Relationships between Tourism Development and Foreign Direct Investment: An Empirical Study in Selected Major Asian Countries. *International Journal of Business Social Science*, *3*(2), pp. 250–257.

Samimi, A. J., and Sadeghi, S. (2013). The Relationship between Foreign Direct Investment and Tourism Development: Evidence from Developing Countries. *International Journal of Institutions and Economies, 5*(2), pp. 59–68.

Selvanathan, S., E. A. Selvanathan, & Viswanathan, B. (2012). Foreign Direct Investment and Tourism: Empirical Evidence from India, Tourism Analysis, 17(1), 91-98.

Hotel Energy Solutions (2011). Analysis of energy use by European hotels: online survey and desk research. Hotel Energy Solutions Project Publication. Available at: https://www.e-unwto.org/doi/pdf/10.18111/9789284415120 (Accessed: 20 February 2021).

Tang, S., E.A. Selvanathan, and Selvanathan, S. (2007). The Relationship between Foreign Direct Investment and Tourism: Empirical Evidence from China. *Tourism Economics*, *13*(1), pp. 25–39.

UNCTC (1992). *The Determinants of Foreign Direct Investment. A Survey of Evidence.* New York: United Nations

UNCTAD (2007). *World Investment Report 2007, FDI in Tourism: The Development Dimension*. New York and Geneva: United Nations.

UNCTAD (2020). *World Investment Report 2020. International Production beyond the Pandemic*. New York and Geneva: United Nations.

UNWTO (2015). UNWTO tourism highlights. Available at: https://www.e-unwto.org/doi/pdf/10.18111/9789284416899 (Accessed: 1 February, 2021).

Vasilevska, I. (1998). Tourism in Macedonia: Review of Development Trends, Characteristics and Possibilities for Future Growth. *Turizam*, *46*(5–6), pp. 269–275.

Virtuoso (2017). Virtuoso's 2018 Luxe report. Available at: https://www.virtuoso.com/getmedia/a66cafd3-866f-4661-b6c3-02f8bfc9a83f/2018-Virtuoso-Luxe-Report-Release_FINAL.aspx (Accessed: 10 May 2021).

Vivid Maps (2019). The over tourism map. Available at: https://vividmaps.com/the-over-tourism-map/ (Accessed: 20 April 2021).

Vorkinek, B. (2020). Overcoming over-tourism. Available at: https://www.cushmanwakefield.com/en/insights/the-edge/overcoming-over-tourism (Accessed: 5 May 2021).

Vukonić, B. (2005). Povijest hrvatskog turizma, Prometej, Zagreb. ISBN 953-6460-45-9

World Travel and Tourism Council (2021). Economic impact reports. Available at: https://wttc.org/Research/Economic-Impact (Accessed: 5 May 2021).

8 Foreign Direct Investment, Tourism Development, and Economic Growth

Evidence from Sub-Saharan Africa

James Antwi

Introduction

The paper examines the causal link between foreign direct investment (FDI), economic growth, and tourism in Sub-Saharan African (SSA) countries. The study investigates the short-run and the long-run effects of FDI and economic growth on the development of the tourism industry using the autoregressive distributed lag (ARDL) model.

Tourism is one of the largest and fastest-growing industries in the world (Mustafa and Santhirasegaram, 2014). It has been an important contributor to the export sector of SSA countries over the last 25 years. International tourist receipts as a percentage of the total exports were 11.2% in 1998 and averaging about 8.1% over the study period.[1] The World Bank statistics show that the annual average growth rate of international tourism arrivals in SSA countries for the years 1995–2019 is about 5.9%. From 2016, Figure 8.1 shows a steady decline of tourism receipt as a percent of export from 9.01% in 2016 to 8.09% in 2019. The maximum and minimum tourism receipts as a percent of exports are 11.12% in 1998 and 5.76% in 2011, respectively. Over the period 1995 to 2019, the total number of tourism arrivals increased from a minimum of 14.24 million in 1995 to a maximum of 56.03 million in 2018 (see Figure 8.2 and Table 8.1). The average number of tourism arrivals is approximately 33.3 million over the last 25 years in SSA.

FDI is a major stimulus to economic growth in developing countries (Owusu-Antwi et al., 2013). According to the authors, FDI brings in needed financial resources, technology, and human capital to grow the economics of the destination countries; as a result, it has received the attention of policymakers in developing countries. Despite the importance of FDI, the flow to SSA countries decreased from about US$45.23 billion in 2015 to nearly US$31.71 billion in 2019, a decrease of approximately 29.9% (see Figure 8.3). Middle African countries experience the worst decline of about 76.7% from 2015 to 2019. West African countries recorded a decrease of 42.6% over the period 2015–2019. However, Southern African countries saw a rise in FDI flow of about 59.5%. Figure 8.3 shows the trend in FDI inward flows for SSA countries over the study period.

DOI: 10.4324/9781003155492-11

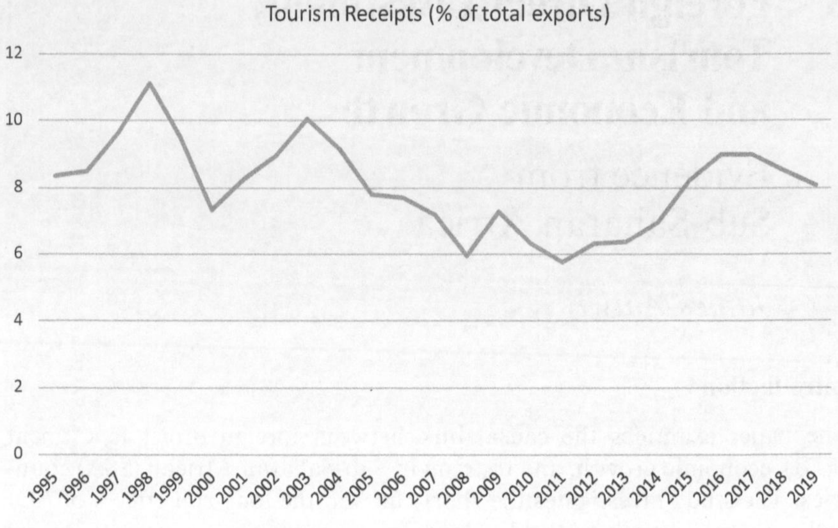

Figure 8.1 Tourism receipts (% of total export) in SSA countries from 1995 to 2019.

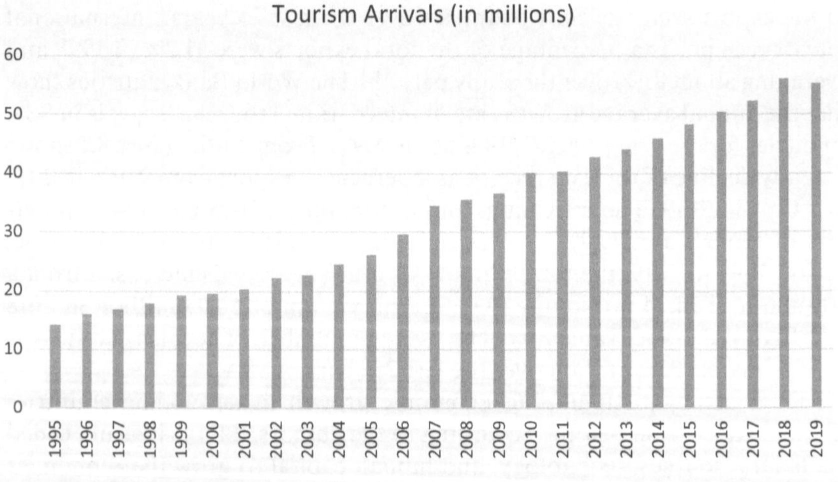

Figure 8.2 International tourism arrivals in SSA countries from 1995 to 2019.

The implication for the decline is that SSA countries need to compete to attract FDI into the sub-region. SSA countries' policymakers must create the enabling environment necessary to bring in FDI. To reverse the declining trend, economic liberalization activities including trade openness necessarily translate into significant FDI inflows into the destination countries. FDI helps by adding to the resources available for investment and capital formation (Owusu-Antwi et al., 2013).

Table 8.1 International tourism for SSA countries

Year	*International tourism, number of arrivals (in millions)*	*International tourism, receipts (current US$ in millions)*	*International tourism, receipts (% of total exports)*
1995	14.24	7,201.60	8.36
1996	16.10	8,010.82	8.50
1997	16.68	8,647.72	9.64
1998	17.80	9,142.86	11.12
1999	19.17	9,509.46	9.47
2000	19.46	8,536.85	7.31
2001	20.39	9,146.05	8.27
2002	22.03	10,135.05	8.95
2003	22.78	14,379.72	10.04
2004	24.55	16,702.40	9.12
2005	26.20	18,958.31	7.81
2006	29.58	21,231.55	7.69
2007	34.45	25,110.38	7.22
2008	35.52	25,423.46	5.97
2009	36.48	23,547.69	7.27
2010	40.09	26,210.21	6.34
2011	40.26	28,933.94	5.76
2012	42.69	31,320.18	6.34
2013	44.20	31,307.51	6.36
2014	45.96	32,471.95	6.96
2015	48.39	29,639.33	8.23
2016	50.61	29,944.07	9.01
2017	52.27	34,084.61	9.00
2018	56.03	36,068.86	8.52
2019	55.25	35,296.56	8.09

Source: World Development Indicators.

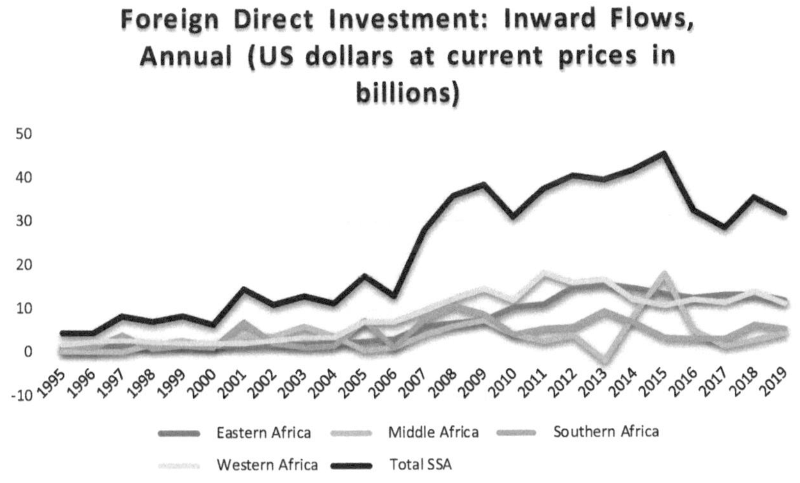

Figure 8.3 The trend in FDI annual flow (US$ at current prices in billions) in SSA countries during 1995–2019.

The evolution of FDI inflows to SSA countries is generally positive from $4.43 billion in 1995 to $31.71 billion in 2019. SSA's FDI annual inward flow reached a maximum of $45.23 billion in 2015 and declined to $31.71 billion in 2019. Similarly, international tourism arrivals have also increased over the study period from 14.24 million tourists in 1995 to 55.25 million tourists in 2019. International tourism receipts as a percent of export marginally decreased from 8.36% in 1995 to 8.09% in 2019 (see Table 8.2).

Empirical studies show that tourism and FDI promote economic growth (Craigwell and Moore, 2008; Read, 2008). According to Tang, Selvanathan, and Selvanathan (2007), FDI plays an important role in developing the tourism industry. Tourism has contributed to economic development in many

Table 8.2 Foreign direct investment: inward flows, annual (US dollars at current prices in billions)

Year	*Eastern Africa*	*Middle Africa*	*Southern Africa*	*Western Africa*	*Total SSA*
1995	0.68	0.36	1.53	1.86	4.43
1996	0.72	0.18	1.07	2.62	4.59
1997	1.24	0.31	4.02	2.72	8.29
1998	1.75	1.83	0.91	2.51	7.00
1999	1.56	2.96	1.69	2.34	8.55
2000	1.47	1.53	1.27	2.13	6.40
2001	1.58	3.72	7.24	2.07	14.61
2002	1.73	3.96	2.28	2.91	10.88
2003	2.02	6.25	1.28	3.36	12.91
2004	2.33	3.70	1.54	3.66	11.23
2005	2.54	0.72	7.29	7.12	17.67
2006	3.33	1.32	1.33	7.05	13.03
2007	6.22	4.30	7.88	9.56	27.96
2008	6.84	6.04	10.57	12.36	35.81
2009	7.49	7.54	8.71	14.76	38.50
2010	10.22	4.32	4.27	12.02	30.83
2011	11.14	2.41	5.52	18.34	37.41
2012	14.82	3.74	5.85	15.92	40.33
2013	15.49	−1.96	9.27	16.57	39.37
2014	14.49	8.46	6.85	12.12	41.92
2015	13.36	17.95	3.21	10.71	45.23
2016	12.42	4.87	2.93	11.96	32.18
2017	12.85	1.20	2.71	11.51	28.27
2018	12.94	2.53	6.06	13.72	35.25
2019	11.54	4.18	5.12	10.87	31.71
Mean*	6.83	3.70	4.42	8.43	23.37
Median*	6.22	3.70	4.02	9.56	27.96
Minimum*	0.68	−1.96	0.91	1.86	4.43
Maximum*	15.49	17.95	10.57	18.34	45.23
Growth rate 2015–2019*	13.62%	−76.71%	59.50%	−42.64%	−29.89%

Source: United Nations Conference on Trade and Development.
*: Author's calculations.

countries (Cortes-Jimenez and Pulina, 2010). Khoshnevis Yazdi, Homa Salehi, and Soheilzad (2017) find a positive relationship between tourism and economic growth. Tourism contributes to foreign exchange income, employment, and economic growth (Craigwell, 2007; Craigwell and Moore, 2008; Read, 2008; Samimi, Sadeghi and Sadeghi, 2017). Studies show a growing interest in the link between FDI and tourism development (Kundu and Contractor, 1999; Sanford Jr and Dong, 2000; Tang et al., 2007; Craigwell and Moore, 2008; Selvanathan, Selvanathan and Viswanathan, 2012; Fereidouni and Al-Mulali, 2014). Despite the importance of FDI and tourism to economic development, to the best of my knowledge, no empirical study has been done on the nature of their linkage for SSA countries using the panel ARDL model.

Prior studies on FDI and tourism have mainly focused on a single country, e.g., Sanford Jr and Dong (2000) for the United States; Balaguer and Cantavella-Jorda (2002) for Spain; Tang et al. (2007) for China; Cortes-Jimenez and Pulina (2010) for Spain and Italy; Selvanathan et al. (2012) for India; Alam et al. (2016) for Saudi Arabia; Khoshnevis Yazdi et al. (2017) for Iran; Yu-Chi and Lin(2018) for Taiwan. Other studies have a focus on regions outside SSA countries, e.g., Khoshnevis Yazdi et al. (2017) for EU. The authors examine the relationship between FDI and tourism development in EU countries; the link between FDI in real estate and tourism in OECD countries (Fereidouni and Al-Mulali, 2014); FDI and tourism in Small Island Developing States or SIDS (Craigwell and Moore, 2008); and FDI and tourism in Asian countries (Salleh, Othman and Sarmidi, 2011).

Some studies have examined the link between tourism and economic growth. Banday and Kocoglu (2015) report tourism as a long-run economic growth factor in an empirical investigation for India using causality analysis; Kreishan (2010) examines tourism and economic growth for Jordan. Mustafa and Santhirasegaram (2014) investigate the relationship between tourism receipts and sustainable economic growth in Sri Lanka, and Oh (2005) shows the contribution of tourism development to economic growth in the Korean economy.

Most of the studies use time series econometric models (Oh, 2005; Tang et al., 2007; Kreishan, 2010; Mustafa and Santhirasegaram, 2014; Banday and Kocoglu, 2015; Alam et al., 2016; Samimi et al., 2017; Yu-Chi and Lin, 2018).

This research builds on prior studies in two important ways. First, this study is the first to use a comprehensive data set from 1995 to 2019 to examine the link between tourism development, FDI, and economic growth for SSA countries using the panel ARDL model. Second, I aim to add more evidence to the causal effect of FDI on the tourism industry for SSA countries.

The structure of the paper is as follows. The next section discusses the relevant literature, focusing mainly on the relationship between foreign direct investment and tourism. The third section explains the methodology, while the fourth section focuses on the empirical analysis and discussion of the results. The fifth section provides conclusions and policy recommendations.

Literature Review

Existing literature shows no consensus on the causal link between FDI and tourism. Tang et al. (2007) examine the relationship between FDI and tourism in China from 1985 and 2001 using a vector autoregressive (VAR) model. The authors report that FDI positively impacts tourism. Samimi et al. (2017) evaluate the causal link between tourism and FDI with data from 1995 to 2008 for some developing countries using the panel Vector Error Correction Model (VECM) model. They record a bilateral long-run causality between FDI and tourism. However, the authors did not find any short-run association between FDI and tourism.

Sanford Jr and Dong (2000) study the association between tourism and FDI in the United States. The writers use the TOBIT model in their study and notice a positive and significant relationship between tourism and FDI in the United States. The authors suggest that international tourism allows potential investors to obtain first-hand knowledge of the environment of the country being visited and, as a result, investment possibilities could be identified (Craigwell and Moore, 2008).

Read (2008) examines the factors that contribute to FDI inflows in SIDS. The results show that openness to trade, income levels, and broad regional location positively impact FDI inflow in SIDS. In their study of the relationship between FDI and tourism in SIDS, Craigwell and Moore (2008) find a bi-directional causal relationship between tourism and FDI inflows. However, the causal relationship mainly runs from FDI to tourism, suggesting that FDI provides the capacity for SIDS countries to grow their tourism industry. Alam et al. (2016) examine the association between FDI and tourism in Saudi Arabia using quarterly time series data from 2000 to 2013. The findings indicate a positive bi-directional causal link between tourism and FDI in the short run and long run.

A 2017 study by Khoshnevis Yazdi, Nateghian and Sheikh Rezaie shows no causal link between FDI and tourist receipts for EU countries. The authors study the effect of FDI, official exchange rate, and trade on tourism development for 27 countries from 1995 to 2014. They employ the pooled mean group model. The evidence suggests no causal association between FDI and tourist receipt. However, the authors find a bi-directional causal link between real exchange rate, trade openness, and tourist receipts.

Yu-Chi and Lin (2018) investigate the existence of causality relationships among international tourist arrivals, foreign exchange income, foreign direct investment, and economic growth (GDP) using Taiwan's tourism data from 1976 to 2016. The results suggest long-run relationships among the variables. The Granger causality results indicate a bi-directional causality between tourism and GDP. The authors find no evidence that tourism causes FDI or FDI causes tourism.

Using a data set from 1960 to 2000, Dritsakis (2004) studied the long-term effects of tourism on the economic growth of Greece. The author used the

VAR model. The result exhibits a bi-directional causality between tourism and economic development. The study also finds a causal relationship between the real exchange rate and tourism. A 2002 study by Balaguer and Cantavella-Jorda also shows that tourism promotes economic growth in Spain over the last three decades. Oh (2005) observes that tourism impacts positively economic growth in South Korea from 1975 to 2001 in the short run. The author uses VAR model in the study. The finding, however, shows no long-run association between tourism and economic growth.

A 2010 study by Kreishan records that tourism promotes economic growth in Jordan in the long run. The author uses time series annual data from 1970 to 2009. Mustafa and Santhirasegaram (2014) investigated long-run and short-run relationships between tourism and economic growth in Sri Lanka. Using data from 1978 to 2011, the findings exhibit a long-run positive tie between tourism and economic growth. According to Banday and Kocoglu (2015), tourism is a major contributor to economic growth in the short run and long run. The evidence is based on time series data from 1991 to 1991 in India using the co-integration test and VECM. Cortes-Jimenez and Pulina (2010) state that tourism causes economic growth for Spain and Italy using data from the 1950s and 1960s, respectively. Table 8.3 shows a summary of literature review on the link between tourism and FDI.

Methodology

The objective of this section is to design an empirical model and estimation techniques to examine the tourism-FDI-economic growth relationship. Table 8.4 shows variable names, description, unit of measurement, and sources. I use international tourism, number of arrivals to proxy tourism. Empirical literature using tourist arrivals to measure tourism includes Willem te Velde and Nair (2006), Tang et al. (2007), Salleh et al. (2011), Fereidouni and Al-Mulali (2014), and Samimi et al. (2017).

I use FDI – annual inward flow to measure FDI; I use real GDP per capita to measure economic growth, and the real effective exchange rate to measure the official exchange rate. Empirical literature using real exchange rate includes Dritsakis (2004) and Khoshnevis Yazdi et al. (2017). I use real

Table 8.3 Summary of literature review

FDI causes tourism	*Tourism causes FDI*	*No relationship*
(Craigwell and Moore, 2008)	(Sanford Jr and Dong, 2000)	(Willem te Velde and Nair, 2006)
(Tang et al., 2007)	(Craigwell and Moore, 2008)	(Khoshnevis Yazdi et al., 2017)
(Samimi et al., 2017)	(Samimi et al., 2017)	(Yu-Chi and Lin, 2018)

Source: Author's compilation.

Table 8.4 Overview of variables

Variable	*Description*	*Unit of measurement*	*Source*
TA	International tourism, number of arrivals	Number in millions	World Development Indicators
FDI	Foreign direct investment: Inward flow, annual	US$ at current prices in billions	United Nations Conference on Trade and Development
GDP	GDP per capita	US$ at constant 2010	World Development Indicators
ExchRate	Real effective exchange rate	Index (2010 = 100)	World Development Indicators

variables based on the availability of data and whenever possible. The real values remove the effect of inflation to obtain a more accurate picture of the trends in the variables.

This research uses a panel of 41 SSA countries over 1995–2019 to examine the relationship between tourism, FDI, and economic growth. Other studies have used panel data set to explain the link between tourism and FDI (Khoshnevis Yazdi et al., 2017). The panel ARDL estimator controls for the endogeneity concerns of FDI and international tourism. Empirical evidence shows the existence of a bi-directional causal relationship between FDI and tourism (Craigwell and Moore, 2008; Fereidouni and Al-Mulali, 2014; Alam et al., 2016; Samimi et al., 2017; Yu-Chi and Lin, 2018).

This paper uses a panel ARDL model and the Pooled Group Mean (PMG) estimator proposed by Pesaran, Shin, and Smith (1999) to examine the effect of FDI and economic growth on tourism development. The data is an unbalanced panel of 41 SSA countries over 1995–2019. Pesaran et al. (1999) use the ARDL (*p*, *q*, …., *q*) model as the empirical structure. The PGM estimator permits the short-run coefficients to vary across countries while restricting the homogeneity of long-term coefficients across countries (Pesaran et al., 1999). The base model for FDI, economic growth, and tourism is as in equation (8.1). The base ARDL (1, 1, 1, 1) model is shown in equation (8.2). The lag order of the ARDL is based on the Akaike Information Criterion (AIC).

$$\ln TA_{it} = \sum_{j=1}^{p} \beta_{ij} \ln TA_{it-j} + \sum_{j=0}^{q} \delta_{ij} FDI_{it-j} + \sum_{j=0}^{q} \phi_{ij} \ln GDP_{it-j} + FE_{i+\varepsilon it} \tag{8.1}$$

$$\ln TA_{it} = \beta_{ij} \ln TA_{it-1} + \delta_{ij} FDI_{it} + \delta_{ij} FDI_{it} + \phi_{i0} \ln GDP_{it} + \phi_{i0} \ln GDP_{it-1} + FE_{i+\varepsilon it} \tag{8.2}$$

where $\ln TA_{it}$ is tourists arrival in year *t* for country *i*, $\ln TA_{it-j}$ is tourists arrival in year *t−j*, FDI_{it} is country *i* FDI in year *t*, FDI_{it-j} is country i FDI in year $t-j$. The $\ln GDP_{it}$ is country i GDP in year t, and $\ln GDP_{it-j}$ is country i GDP in year $t-j$, FE_i is the unobserved country fixed effects, and ε_{it} is an error.

The re-parametrized ARDL (1, 1, 1, 1) error correction model is specified in equation (8.4), which is derived by adding equations (8.2) and (8.3) and rearranging it.

$$-\ln \mathrm{TA}_{\mathrm{it-1}} = \ln \mathrm{TA}_{\mathrm{it-1}} + \delta_{\mathrm{it}} + \phi_{\mathrm{it}} - \phi_{\mathrm{i1}} - \phi_{\mathrm{i1}} \ln \mathrm{GDP}_{\mathrm{it}} \tag{8.3}$$

$$\Delta \ln \mathrm{TA}_{it} = \theta \left[\ln TA_{it-1} - \lambda'_i X_{it} \right] + \Gamma'_i \Delta X_{it} \tag{8.4}$$

Notes:

- $\theta_i = -(1-\beta_i)$ is the group-specific speed of adjustment coefficient (expected $-2 < \theta_i < 0$).
- $\lambda'_i = \left[\left(\dfrac{\delta_{i0} + \delta_{i1}}{1-\beta_i} \right) \left(\dfrac{\phi_{i0} + \phi_{i1}}{1-\beta_i} \right) \right]$ is a vector of long-run relationships.
- $X'_{it} = [FDI_{it}\ \mathrm{InGSDP}_{it}]$
- $[\mathrm{In}TA_{it-1} - \mathrm{In}\,\mathrm{GSDP}_{it}]$ is the error correction term.
- $\Gamma'_i = -[\delta_{i1}\ \phi_{i1}]$ is a vector of the short-run dynamic coefficients.
- Δ is the first difference operator.

Another estimator, the mean group (MG), is consistent when the slope and intercepts differ across countries (Pesaran et al., 1999). I use the Hausman test to select between the PMG and the MG estimators based on their consistency and efficiency properties (Pesaran et al., 1999). The PMG estimates are consistent and efficient in a long-run link between stationary (i.e., I(0)) and integrated (i.e., I(1)) variables (Pesaran et al., 1999). For the PMG estimates to be consistent and efficient, the coefficient on the error-correction term should be negative and greater than −2 (Loayza and Ranciere, 2004).

Empirical Results

This section presents the results of the link between tourism, FDI, and economic growth. The results of estimating equation (8.4) are presented in Tables 8.6 and 8.7. Table 8.5 shows descriptive statistics. The ARDL model does not require pre-testing of the order of integration of the variables (Lee and Wang, 2015). Tables 8.6 and 8.7 show the coefficient of the error correction term between −0.350 and −0.259 and are significant at the 1% level. They also have the required negative sign, suggesting the speed of adjustment toward the long run is between 25.9% and 35%. Table 8.6 shows the Hausman test and the long-run and short-run coefficients for economic growth and FDI. The Hausman test implies that PGM estimation is appropriate.

Table 8.5 Summary statistics

Variable	*Obs.*	*Mean*	*Std. Dev.*	*Minimum*	*Maximum*
Dependent Variables					
Log tourist arrivals	838	12.604	1.565	7.973	16.532
Explanatory Variables					
Foreign direct investment	838	7.124	1.914	0.280	12.098
Log gross domestic product	838	7.002	0.959	5.213	9.619

Table 8.6 The long- and short-run effects of foreign direct investment (FDI) on tourism *dependent variable: log tourist arrivals dynamic specification: ARDL (1 1 1)*

Estimator	*PMG*	*MG*	*Hausman test*	*Robust OLS*
Long-run coefficients				
Log foreign direct investment	0.143*** (0.015)	1.450 (1.095)	H_0: PMG H_a: MG	0.469*** (0.021)
Log gross domestic product	1.431*** (0.087)	−3.967 (4.564)	Chi-square statistic = 0.65 [0.723]	0.40*** (0.039)
Error correction coefficients: Phi	−0.350*** (0.093)	−0.588*** (0.069)		
Short-run coefficients				
Δ Log foreign direct investment	−0.061 (0.098)	0.002 (0.093)		
Δ Log gross domestic product	0.876** (0.301)	0.827 (0.589)		
Constant	0.586*** (0.133)	−0.012 (2.523)		6.493*** (0.284)
F-statistics				374.06***
R-squared				0.484
Number of countries	41			41
Number of observations	772			838

Note: ***, **, * significant at 1%, 5% or 10%, respectively. Standard errors in parentheses.

Findings from Table 8.6 means that FDI and economic growth positively impact tourism development in the long run. The result implies that in the long run, FDI and economic growth are vital for tourism growth in SSA. Table 8.6 also shows a short-run positive effect of economic growth on tourism.

According to Table 8.6, FDI and economic growth are statistically significant at a 1% significant level. In the long run, the coefficient of FDI and economic growth are 0.143 and 1.431, respectively. This means that a 1% increase in FDI and economic growth will lead to a 0.143% and 1.431% increase in tourism, respectively. In the short run, the coefficient of economic

Table 8.7 The Long- and Short-Run Effects of Foreign Direct Investment (FDI) on Tourism *Dependent Variable: Log Tourist Arrivals Dynamic specification: ARDL (1 1 1 1)*

Estimator	*PMG*			*Robust OLS*		
Model	*1*	*2*	*3*	*4*	*5*	*6*
Long-run coefficients						
Log foreign direct investment	0.216***	0.143***	0.474***	0.538***	0.469***	0.268***
	(0.016)	(0.015)	(9.092)	(0.022)	(0.021)	(0.047)
Log GDP		1.431***	0.711*		0.40***	0.793***
		(0.087)	(0.367)		(0.039)	(0.114)
Official exchange rate			−0.003*			−0.002
			(0.002)			(0.002)
Error correction coefficients: Phi	−0.276***	−0.350***	−0.259***			
	(0.039)	(0.093)	(0.057)			
Short-run coefficients						
Δ Foreign direct investment	−0.037	−0.061	0.0723			
	(0.095)	(0.098)	(0.094)			
Δ Log GDP		0.876**	1.427**			
		(0.301)	(0.723)			
Δ Official exchange rate			−0.002			
			(0.003)			
constant	3.039***	0.586***	1.341***	8.775***	6.493***	5.824***
	(0.415)	(0.133)	(0.298)	(0.159)	(0.284)	(0.584)
F-statistics	N/A	N/A	N/A	652.18***	374.06***	224.54***
R-squared	N/A	N/A	N/A	0.432	0.484	0.715
Number of countries	41	41	13	41	41	13
Number of observations	772	772	250	838	838	272

Note: ***, **, * significant at 1%, 5% or 10%, respectively. Standard errors in parentheses.

growth is 0.876 and it is significant at the 5% significant level. A 1% increase in economic growth will result in a 0.876% rise in tourism in the short run. Also, the results of a robust OLS estimation show a strong correlation between FDI and tourism as well as between economic growth and tourism.

Table 8.7 shows the results of estimating equation (8.4) for FDI, economic growth, and tourism. I estimate six models that include model 1, which contains only FDI as an explanatory variable. Model 2 adds economic growth, and Model 3 adds real exchange rate to the equation. I also run three robust OLS models. For all the six models, the coefficients of FDI are statistically significant at a 1% level. Models 1, 2, and 3 show that FDI positively impacts tourism in the long run. Moreover, models 4, 5, and 6 suggest a strong correlation between FDI and tourism. The findings are in line with the works of Tang et al. (2007), Craigwell and Moore (2008), and Samimi et al. (2017). The positive link between tourism and GDP suggests that richer countries attract tourists as compared to poor countries. This may be that richer countries are better able to provide better infrastructure including hotels and better security that attract international tourist arrivals.

The results of this research are generally consistent with other empirical studies on the link between tourism, FDI, and economic growth. The evidence suggests that FDI and economic growth are important drivers of tourism in the long run. Also, economic growth positively affects tourism in the short run. However, the finding of Table 8.7 shows that exchange rate appreciation increases the price of goods and services and therefore causes a fall in export including tourism, confirming the theoretical prediction of exchange rate appreciation on export.

Conclusion and Policy Recommendation

Tourism has been an important contributor to the export sector of SSA countries over the last 25 years. International tourist receipts as a percentage of the total exports were 11.2% in 1998 and averaging about 8.1% over the study period. Tourism receipts as a percent of export have fallen from 9.01% in 2016 to 8.09% in 2019. Over the period 1995 to 2019, the total number of tourism arrivals increased from a minimum of 14.24 million in 1995 to a maximum of 56.03 million in 2018. The average number of tourism arrivals is approximately 33.3 million over the last 25 years in SSA countries.

FDI is a major stimulus to economic growth in developing countries attracting foreign exchange, technology, and human capital to grow the economics of the destination countries. As a result, it has received the attention of policymakers in developing countries. Despite the importance of FDI, the flow to SSA countries decreased from about US$45.23 billion in 2015 to nearly US$31.71 billion in 2019, a decrease of approximately 29.9%. The implication for the decline is that SSA countries need to compete to attract FDI into the sub-region. SSA countries' policymakers must create the enabling environment necessary to bring in FDI. To reverse the declining trend, economic liberalization activities including trade openness are necessary to translate into significant FDI inflows into SSA countries.

Empirical evidence shows that FDI and tourism are necessary for economic growth. This study uses a panel data set of 41 SSA countries from 1995 to 2019. The results of the ARDL model suggest that FDI and economic growth are key drivers of tourism growth in the long run in SSA countries. The findings are consistent with the works of Tang et al. (2007), Craigwell and Moore (2008), and Samimi et al. (2017). Policymakers should implement programs that promote FDI with the attendant results of tourism development and economic growth.

Note

1 The World Bank: World Development Indicators.

References

Alam, A. et al. (2016) 'The relationship between tourism, foreign direct investment and economic growth: evidence from Saudi Arabia', *European Academic Research*, 4(4), pp. 4091–4106.

Balaguer, J. and Cantavella-Jorda, M. (2002) 'Tourism as a long-run economic growth factor: the Spanish case', *Applied Economics*, 34(7), pp. 877–884.

Banday, U. J. and Kocoglu, M. (2015) 'Tourism as a long-run economic growth factor: an empirical investigation for India using causality analysis', *World Journal of Islamic History and Civilization*, 5(2), pp. 48–53.

Cortes-Jimenez, I. and Pulina, M. (2010) 'Inbound tourism and long-run economic growth', *Current Issues in Tourism*, 13(1), pp. 61–74.

Craigwell, R. (2007) 'Tourism competitiveness in small island developing states. Research paper No 2007/19', World Institute for Development Economics Research, Helsinki.

Craigwell, R. and Moore, W. (2008) 'Foreign direct investment and tourism in SIDS: Evidence from panel causality tests', *Tourism Analysis*, 13(4), pp. 427–432.

Dritsakis, N. (2004) 'Tourism as a long-run economic growth factor: an empirical investigation for Greece using causality analysis', *Tourism Economics*, 10(3), pp. 305–316.

Fereidouni, H. G. and Al-Mulali, U. (2014) 'The interaction between tourism and FDI in real estate in OECD countries', *Current Issues in Tourism*, 17(2), pp. 105–113.

Khoshnevis Yazdi, S., Homa Salehi, K. and Soheilzad, M. (2017) 'The relationship between tourism, foreign direct investment and economic growth: evidence from Iran', *Current Issues in Tourism*, 20(1), pp. 15–26.

Khoshnevis Yazdi, S., Nateghian, N. and Sheikh Rezaie, N. (2017) 'The causality relationships between tourism development and foreign direct investment: an empirical study in EU countries', *Journal of Policy Research in Tourism, Leisure and Events*, 9(3), pp. 247–263.

Kreishan, F. M. (2010) 'Tourism and economic growth: The case of Jordan', *European Journal of Social Sciences*, 15(2), pp. 63–68.

Kundu, S. K. and Contractor, F. J. (1999) 'Country location choices of service multinationals: An empirical study of the international hotel sector', *Journal of International Management*, 5(4), pp. 299–317.

Lee, Y.-M. and Wang, K.-M. (2015) 'Dynamic heterogeneous panel analysis of the correlation between stock prices and exchange rates', *Economic research-Ekonomska istraživanja*, 28(1), pp. 749–772.

Loayza, N. and Ranciere, R. (2004) *Financial development, financial fragility, and growth*. The World Bank.

Mustafa, A. and Santhirasegaram, S. (2014) 'Empirical investigation of the relationship between tourism receipts and sustainable economic growth in Sri Lanka', *Journal of Emerging Trends in Economics and Management Sciences*, 5(7), pp. 131–137.

Oh, C.-O. (2005) 'The contribution of tourism development to economic growth in the Korean economy', *Tourism Management*, 26(1), pp. 39–44.

Owusu-Antwi, G., Antwi, J. and Poku, P. K. (2013) 'Foreign direct investment: a journey to economic growth in Ghana-empirical evidence', *International Business & Economics Research Journal (IBER)*, 12(5), pp. 573–584.

Pesaran, M. H., Shin, Y. and Smith, R. P. (1999) 'Pooled mean group estimation of dynamic heterogeneous panels', *Journal of the American statistical Association*, 94(446), pp. 621–634.

Read, R. (2008) 'Foreign direct investment in small island developing states', *Journal of International Development: The Journal of the Development Studies Association*, 20(4), pp. 502–525.

Salleh, N. H. M., Othman, R. and Sarmidi, T. (2011) 'An analysis of the relationships between tourism development and foreign direct investment: an empirical study in elected major Asian countries', *International Journal of Business and Social Science*, 2(17), pp. 250–257.

Samimi, A. J., Sadeghi, S. and Sadeghi, S. (2017) 'The relationship between foreign direct investment and tourism development: evidence from developing countries', *Institutions and Economies*, 5(2), pp. 59–68.

Sanford Jr, D. M. and Dong, H. (2000) 'Investment in familiar territory: Tourism and new foreign direct investment', *Tourism Economics*, 6(3), pp. 205–219.

Selvanathan, S., Selvanathan, E. A. and Viswanathan, B. (2012) 'Causality between foreign direct investment and tourism: empirical evidence from India', *Tourism Analysis*, 17(1), pp. 91–98.

Tang, S., Selvanathan, E. A. and Selvanathan, S. (2007) 'The relationship between foreign direct investment and tourism: empirical evidence from China', *Tourism Economics*, 13(1), pp. 25–39.

Willem te Velde, D. and Nair, S. (2006) 'Foreign direct investment, services trade negotiations and development: the case of tourism in the Caribbean', *Development Policy Review*, 24(4), pp. 437–454.

Yu-Chi, S. and Lin, H.-P. (2018) 'Causality relationship between tourism, foreign direct investment and economic growth in Taiwan', *Asian Journal of Economic Modelling*, 6(3), pp. 287–293.

9 The Relationship between FDI and Tourism Growth

A Case Study of China

Eliyathamby A Selvanathan, Maneka Jayasinghe and Saroja Selvanathan

Introduction

Tourism is one of the fastest growing industries and plays a significant role in the global social and economic arena. According to the World Tourism Organization (WTO), in 2019, the tourism sector generated US$1,733 billion in export revenue, which is around 7% of the world exports of goods and services. During the same year, 1,462 million tourists travelled around the world. With globalization, tourism has become a major export sector in many countries. While the tourism sector generates income, taxes, foreign exchange earnings and employment (Choi & Sirakaya, 2006; Dwyer & Forsyth, 2008; Jayasinghe & Selvanathan, 2021), it also creates the same multiplier effect on a number of other sectors of the economy, such as transport and communication, construction, education, health, retail and services, eventually leading to increased overall economic growth in most countries. One of the key elements for a successful tourism sector in a country is having good quality tourism related infrastructure, such as hotels, restaurants, transport, communication, highways etc. Most countries, especially developing countries, do not have the resources to invest in tourism-related infrastructure and hence they heavily depend on Foreign Direct Investments (FDI). Therefore, naturally, FDI plays a significant role in the growth of the tourism sector.

When China started to open its economy to the outside world in 1979 there was almost no FDI inflow. However, this rapidly increased to US$235.4 billion in 2018 (1.7% of China's GDP), with FDI stock totalling US$3,557 billion in 2019 over the period 1979 to 2019. During the initial period, most of the FDI went into manufacturing. However, from the 1980s, when the Chinese government introduced tax exemption policies on hotels constructed by foreign investors, FDI received into China by the tourism sector slowly started to gain momentum. This has resulted in China becoming the largest FDI recipient over the last two centuries (Das, 2007).

The tourism industry in China has undergone rapid development since 1978 and at present it is considered as one of the world's top 10 tourism destinations (Arain et al., 2019; Tsang & Hsu, 2011). For example, in 2019 China was

DOI: 10.4324/9781003155492-12

ranked as the fourth highest tourist destination, receiving 145 million international tourist arrivals, compatriots from Hong Kong, Macao and Taiwan, and overnight visitors (Chinese Statistical Yearbook, 2019). Consequently, during the past three decades, the tourism and hospitality industry in China has developed from being small to large, and from rapid growth to maturity (Tsang & Hsu, 2011; Tisdell & Wen, 1991). A detailed analysis of the evolution of the Chinese tourism and hospitality industry can be found in Tsang and Hsu (2011). While the relationship between FDI and tourism is quite intuitive, still there is a dearth of literature on this topic with respect to developing countries, and in particular about China (for example, see, Tang et al., 2007).

The objective of this study is to examine the relationship between FDI and tourism in China. For this purpose, we use two indicators to measure tourism in China; international tourist arrivals and international tourism receipts. While there is an emerging body of literature on this topic, some authors have highlighted that published research which examined the empirical relationship between international tourism and FDI is quite rare and hence the need for further research (Katircioglu, 2011; Samimi et al., 2017). To this end, the current study expands the existing literature on this topic. This study differs from existing research on the link between FDI and tourism in China in two ways. First, the current study uses the most recent data available (up to 2019) for the empirical analysis (see for example, Tang et al., 2007). Second, it uses two indicators to measure tourism – tourist arrivals and tourism receipts (see for example, Tang et al., 2007; Chen, 2017). This study provides a comprehensive and up-to-date analysis of the relationship between FDI and tourism.

The organization of the paper is as follows. Section 'Review of Literature' presents a review of the literature on tourism and FDI. The following section discusses the data used in the paper and presents a preliminary analysis of the data. 'Modelling the Relationship between Tourist Arrivals, Tourism Receipts and FDI' section presents the methodology to be used to model the relationship between FDI and tourism and presents an econometric estimation of the relationship. Finally, in section 'Policy Implications and Concluding Comments', we discuss policy implications and present the concluding comments.

Review of Literature

There is a growing body of literature on the relationship between tourism activity and FDI, but publications are still scarce (Katircioglu, 2011; Samimi et al., 2017). While some of the existing studies focus on the relationship between tourism-related FDI, such as hotels and restaurants and tourism growth, other studies focus on general FDI inflows and tourism growth. In all of these studies, the number of international tourist arrivals has been used as the indicator of tourism. Below we discuss the literature, with a special emphasis on the literature regarding China.

There are a number of studies that investigated the link between tourism and FDI with respect to the hotel industry (see, for example, Dunning and McQueen, 1981; Kundu and Contractor, 1999). Overall, these studies revealed that rates of tourism growth, particularly business tourism, are important determinants of FDI in international hotels. A majority of early research on FDI and tourism was either based on qualitative research or used a single equation approach. In addition, these studies did not focus on examining the causal relationship between FDI and tourism. To overcome this gap in the literature, some studies have highlighted the need for research based on complete simultaneous equations models (see, for example, Gupta, 1983; Witt & Witt, 1995; Lim, 1997; Shan & Wilson, 2001).

Sanford and Dong (2000) investigated the effect of tourism on FDI. The study argued that incorporating tourism improves on existing research on FDI. Empirical analysis based on a TOBIT model estimation showed a positive and significant relationship between tourism and subsequent new FDI in the USA. The analysis, however, did not support the industry-specific effects, which suggest that tourism is associated with increased investment in capital-intensive as well as service industries. The study does not investigate the causal relationship between the variables of interest. Garcia-Flores et al. (2008) examined the relationship between FDI, tourism development and its impacts on the environment in Mexico from 1982 to 2007. The findings showed that there is a positive relationship between FDI and tourism development. Katircioglu (2011) employed the bounds test approach, Granger causality tests and vector error correction modelling to examine the long run equilibrium relationship between inbound tourism and FDI inflows in Turkey between 1970 and 2005. Results suggested that both variables are in a long-run equilibrium relationship only when FDI is a dependent variable under the Autoregressive Distributed Lag (ARDL) modelling approach. The results revealed that international tourism is a catalyst for FDI in the long run in Turkey. The study emphasized the need for Turkish authorities to promote international tourism as FDI and thereby the economy would be positively influenced by the development of this 'foreign exchange' earning sector. Using the quarterly data for the period 1995:2 to 2007:2, Selvanathan et al. (2012) investigated the causal relationship between FDI and tourist arrivals in India. The study employed various time series econometric techniques, such as unit root test and cointegration. The study also employed the Vector Autoregressive (VAR) system in first-difference of the two variables to investigate the causality between tourist arrivals and FDI. The results revealed that there is only a one-way causal relationship from FDI to tourism in India. Samimi et al. (2017) investigated the existence of Granger causality and cointegrated relationships between tourism related FDI and tourist arrivals in 20 developing countries using panel Vector Error Correction Model (VECM) techniques from 1995 to 2008. The results indicated the existence of a cointegrated relationship between variables in the long run. In addition, the study found that there is a bidirectional long-run causality between tourism-related FDI and tourist arrivals and that there is no short-run causality between variables. Bezić and Nikšić Radić (2017) investigated the causal relationship between FDI in the tourism sector and tourism gross value added in Croatia, using quarterly time series data from 2000:1 to 2012:4. The study employed various econometric techniques, such as the unit root test, Johansen cointegration and the Granger causality test, in a VECM, and the Toda–Yamamoto causality test in a VAR Model. The results indicated the existence of a stable cointegrated relationship between variables in the long term. A short-term relationship was also proven between FDI in tourism and gross value added, using the Toda–Yamamoto causality test. Additionally, the study also found a causality between FDI and tourism gross value added.

Amin et al. (2020) examined the relationship between FDI and tourism receipts in Bangladesh using the annual time series data from 1972 to 2017. The study employed various econometric techniques, such as the Augmented Dickey-Fuller (ADF), Phillips-Perron (PP) and Zivot Andrews structural break augmented unit root tests, Johansen cointegration test, Granger causality test, VECM, Dynamic Ordinary Least Square (DOLS) and the ARDL estimation methods and Cumulative Sum (CUSUM) stability test for the analysis. The results revealed that there is a unidirectional causal relationship running from FDI to tourism in the short run and long run in the case of Bangladesh. The long-run estimation results of DOLS indicated that a 1 million USD increase in FDI will increase tourism receipts by 0.065 million USD, while ARDL results indicated that a 1 million USD increase in FDI will increase tourism receipts by 0.062 million USD, respectively.

There are several studies that examine the relationship between FDI and tourism in China. Tisdell and Wen (1991), in a qualitative study, discussed various policy issues with respect to FDI inflows in the tourism industry in China during the period 1979–1988. The study revealed that the fast growth rate of the tourism sector in China is one of the driving forces of foreign investment in China's tourism industry. Tang et al. (2007) investigated the causal link between FDI and tourism in China, using the Granger causality test under a VAR framework. For this purpose, the study used quarterly time series data for the period 1985:1 to 2001:3. The results showed a unidirectional causality running from FDI to tourism. Chen (2010) analysed the influence of FDI within China's tourism industry considering the imbalance of the development process across coastal and inland regions from 1978 to 2008. The results demonstrated that impacts of FDI on the tourism industry in the coastal regions are greater than those of inland regions. The study concludes that the coastal regions have experienced rapid economic and tourism development followed by the inflow of FDI and political preferences.

Chen (2017) examined the relationship between inbound tourism and FDI in China over the period 2000–2014, using the dynamic panel approach.

The empirical results showed that growth of inbound tourism promotes FDI into the tourism industries as well as to other sectors in the economy. The study concluded that the flourishing inbound tourism may have spillover effects on FDI into non-tourism sectors. The study also revealed that the inbound tourism development policy and inward FDI promotion policy are complementary, and hence improved coordination can maximize the effect of policies. Arain et al. (2019) investigated the asymmetric effect of inbound tourism on FDI in the world's top tourist destinations (China, Russia, USA, Spain, Mexico, Italy, Germany, France, Turkey, and the UK), using monthly data over the period 1995 and 2017. The study employed the quantile-on-quantile (QQ) approach as it assesses how various quantiles of inbound tourism affect different quantiles of FDI. Additionally, the Granger causality test in quantiles was also employed in this study to investigate the causal relationship between tourism and FDI in selected countries. The empirical results indicated that the relationship between inbound tourism and FDI is mostly positive for all countries except Mexico and Russia on low and middle quantiles. Nevertheless, significant differences between countries and across all quantiles of inbound tourism and FDI were found. Applying the negative binomial model to 2001–2015 panel data from China, Song et al. (2019) examined the spillover effect of inward FDI on tourism-based outward FDI. The results confirmed the positive effect of China's inward FDI on its tourism-based outward FDI. The study also revealed that both the FDI penetration and tourism-based inward FDI in a province were found to significantly affect the province's tourism-based outward FDI. In particular, foreign investments in highly rated hotels substantially affected tourism-based outward FDI.

In summary, a majority of studies find that there is a positive relationship between FDI and tourism development. However, the direction of the causality varies depending on the country of investigation, econometric techniques used and the period of data used in the analysis. The next section describes the data on China used in the current study and presents a preliminary data analysis.

Preliminary Data Analysis

In this paper, we use data on three variables, namely number of international tourist arrivals (TA), international tourism receipts (TR) and Foreign Direct Investment (FDI) inflows into China. We use data for the period 1978–2019 for the above three variables. The data for these variables are collected from various issues of the China Statistical Yearbook.

Table 9.1 presents the original level form data and the respective growth rates for TA, TR and FDI for selected years during the sample period 1978–2019. Figure 9.1 A–G plot these data in graphical form. As can be seen, the number of international tourist arrivals in China was below 50 million until 1995 which nearly tripled to 133 million within the next 15 years and

Table 9.1 Descriptive summary statistics

	International tourist arrivals		***International tourism receipts***		***FDI inflows to China***		
Year (1)	*Millions (2)*	*Growth rate (3)*	*US$ million (4)*	*Growth rate (5)*	*US$ billion (6)*	*Growth rate (7)*	*% of GDP (8)*
1980	5.7	35.7	0.6	50.0	0.1		0.0
1985	17.8	38.0	1.3	18.2	1.7	31.9	0.5
1990	27.5	12.2	2.2	15.8	3.5	2.8	1.0
1995	46.4	6.2	8.7	19.2	35.8	6.1	4.9
2000	83.4	14.6	17.3	15.4	42.1	2.6	3.5
2005	120.3	10.3	29.3	5.6	104.1	52.8	4.6
2010	133.8	5.8	45.8	15.5	243.7	86.0	4.0
2015	133.8	4.1	45.0	2.1	242.5	−9.6	2.2
2019	145.0	2.7	35.8	−11.3	155.8	−33.8	1.1
Mean	74.1	13.0	19.6	13.6	86.8	32.4	2.6

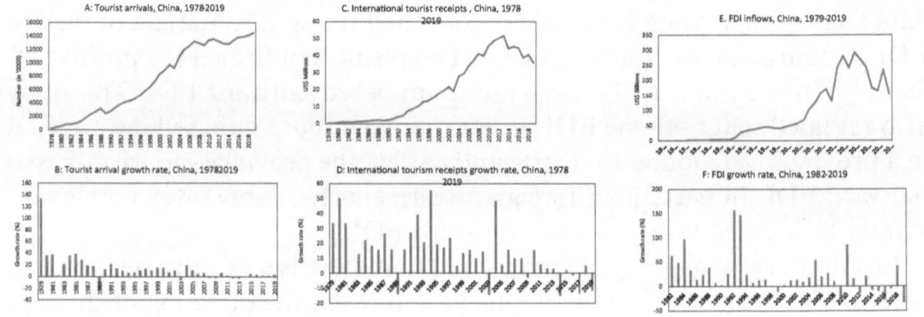

Figure 9.1 International tourist arrivals, tourism receipts and FDI, and their growth rates, China, 1978–2019.

then reached 145 million in 2019, just before the onset of the COVID-19 pandemic.

Generally, an upward trend can be seen in Figure 9.1-A over the sample period of four decades, 1978–2019, with some falls during 1989 (Tiananmen Square incident), 2003 (SARS virus), 2008–2009 (Anti-China protests by Pro-Tibet activist, Swine flu and the Global Financial Crisis), 2012–2014 (start of Hong Kong Protests). This is also reflected in Figure 9.1-B where we plot the growth rate of tourist arrivals. As can be seen from Figure 9.1-C, the international tourism receipts to China has shown a similar pattern to the number of tourist arrivals while there is a downward trend in the receipts in the later years 2014–2019. This could be partly due to the Hong Kong protests and also due to the trade dispute between Beijing and Washington that have been adversely affecting the tourism industry in China

(Kurtenbach and Oslon, 2020). This observation is also reflected in Figure 9.1-D where we plot the growth rate in international tourism receipts to China. Figure 9.1-E and F plots the FDI inflows to China and its growth rate during the sample period 1978–2019. As can be seen, while almost nil in 1978, the FDI inflow is small until 1991, reached double digit numbers in 1992 and increased to a three-digit figure in 2005. FDI reached a maximum of US$291 billion in 2013 and started to fall from 2014. Figure 9.1-G plots the FDI inflows as a percentage of GDP. As can be seen, FDI inflows as a percentage of GDP shows a positive trend until 1993 and then declines steadily (Figure 9.2).

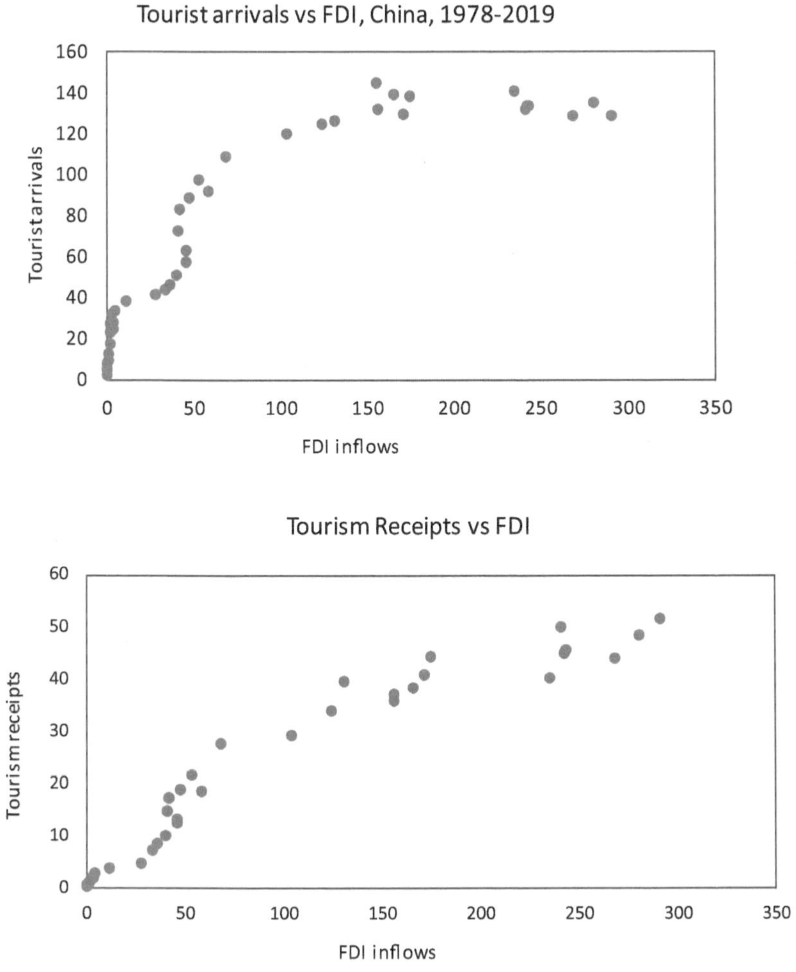

Figure 9.2 Tourist arrivals and tourism receipts vs FDI, China, 1978–2019.

Modelling the Relationship between Tourist Arrivals, Tourism Receipts and FDI

In this section we address a number issues in relation to the relationship between (1) international tourist arrivals (TA) and FDI and (2) International tourisms receipts (TR) and FDI. In constructing these two relationships first we need to consider the causality between TA and FDI and TR and FDI. We employ the Granger (1986) causality test to investigate these two causal relationships between the two set of variables. For example, to test the Granger causality between TA and FDI, we specify the following bivariate VAR system of order *p* for the two variables TA and FDI (for example, see Zapata and Rambaldi, 1997; Gujarati, 1995)

$$TA_t = \lambda_{01} + \sum_{i=1}^{p} \alpha_{i1} TA_{t-i} + \sum_{i=1}^{p} \beta_{i1} FDI_{t-i} + \varepsilon_{1t} \tag{9.1}$$

$$FDI_t = \lambda_{02} + \sum_{i=1}^{p} \alpha_{i2} FDI_{t-i} + \sum_{i=1}^{p} \beta_{i2} TA_{t-i} + \varepsilon_{2t} \tag{9.2}$$

The null hypothesis of 'FDI does not Granger cause TA' can be tested by using standard F-test of the joint hypothesis

$$\mathrm{H}_0 : \beta_{11} = \beta_{21} = \ldots \beta_{p1} = 0 \tag{9.3}$$

If the above null hypothesis is rejected, then we conclude that 'FDI Granger causes TA'. Similarly, if the following null hypothesis is rejected we conclude that 'TA Granger causes FDI'.

$$\mathrm{H}_0 : \beta_{12} = \beta_{22} = \ldots \beta_{p2} = 0 \tag{9.4}$$

Table 9.2 presents the causality test results. As can be seen, at the 5% level, TA Granger causes FDI and at the 10% level FDI Granger causes TA. Similarly, at the 5% level, FDI Granger causes TR and TR Granger causes FDI. This means that there is a two-way causality between TA and FDI as well as TR and FDI.

The next step is to investigate the relationship between TA and FDI in the short run and in the long run. For this purpose, we use the ARDL model and the bounds test based on ARDL. Based on the causality results above, we can set up the long-run regression models for estimation in the form

$$\log(TA) = \alpha + \beta \log(FDI) + \varepsilon \tag{9.5}$$

and

$$\log(FDI) = \alpha + \beta \log(TA) + \varepsilon \tag{9.6}$$

Table 9.2 Pairwise Granger causality tests

Null hypothesis (1)	*F-statistic* (2)	*p-value* (3)
FDI does not Granger Cause TA	2.71	0.08
TA does not Granger Cause FDI	3.95	0.03
FDI does not Granger Cause TR	4.98	0.01
TR does not Granger Cause FDI	9.39	0.00

The attraction of the ARDL bounds test approach is that, (a) it does not require all the variables of the model to be integrated of the same order, that is, all I(0) or all I(1); (b) it can be used in small sample situations as well; and (c) long-run and short-run models can be estimated simultaneously. The ARDL(p, q) formulations of equation (9.5) and equation (9.6) can be written as

$$\Delta(\ln TA_t) = \alpha_1 + \beta_1 \ln TA_t + \beta_2 \ln FDI_t + \sum_{j=1}^{p} \gamma_{1j}\, \Delta\left(\ln TA_{t-j}\right) + \sum_{j=0}^{q} \gamma_{2j}\, \Delta\left(\ln FDI_{t-j}\right) + u_t \quad (9.7)$$

$$\Delta(\ln \mathrm{FDI}_t) = \alpha_1 + \beta_1 \ln \mathrm{FDI}_t + \beta_2 \ln \mathrm{TA}_t + \sum_{j=1}^{p} \gamma_{1j}\, \Delta\left(\ln \mathrm{FDI}_{t-j}\right) + \sum_{j=0}^{q} \gamma_{2j}\, \Delta\left(\ln \mathrm{TA}_{t-j}\right) + v_t \quad (9.8)$$

The ARDL bounds test can be used to test the long-run (or the cointegrating) relationship between TA and FDI based on a non-standard F-test by testing the null and the alternative hypotheses;

$$H_0 : \beta_1 = \beta_2 = 0 \text{ (no cointegration)}$$
$$H_1 : \text{At least on of } \beta_1 \text{ or } \beta_2 \neq 0 \text{ (existance of cointegration)}$$

using the critical values provided in Pesaran et al. (2001). There are two sets of critical values based on the assumptions that all variables are I(0) and the other is that all variables are I(1). If the calculated F-statistic is above the I(1) critical value, then the null hypothesis should be rejected and conclude that there is support for cointegration. If the calculated F-statistic is below the I(0) critical value then the null hypothesis should not be rejected and conclude that there is no support for cointegration. If the calculated F-statistic is between the I(0) critical value and the I(1) critical value, then the test result is inconclusive.

If we find cointegration between TA and FDI, then we can use an error correction model to estimate the speed of adjustment of the disequilibrium caused by previous period shocks that re-converges to the long-run equilibrium. The error correction models which correspond to equations (9.7) and (9.8) can be written as

$$\Delta(\ln TA_t) = \alpha_1 + \sum_{j=1}^{p} \gamma_{1j}\Delta(\ln TA_{t-j}) + \sum_{j=0}^{q} \gamma_{2j}\Delta(\ln FDI_{t-j}) + \delta EC_{t-1} + u_t \tag{9.9}$$

$$\Delta(\ln FDI_t) = \alpha_1 + \sum_{j=1}^{p} \gamma_{1j}\Delta(\ln FDI_{t-j}) + \sum_{j=0}^{q} \gamma_{2j}\Delta(\ln TA_{t-j}) + \delta EC_{t-1} + v_t \tag{9.10}$$

where EC_{t-1} is the error correction term derived from the long-run relationship (9.5) for equation (9.9) and from the long-run relationship (9.6) for equation (9.10). The coefficients γ_{ij} measure the short-run dynamics associated with the long-run relationships and coefficient δ is the speed of adjustment.

Using the error correction models (9.9) and (9.10) we can also test three forms of Granger causality, namely (a) short run causality by performing an F-test on the short-run γ_{ij} coefficients; (b) a long-run causality using a t-test on the error correction coefficient, δ; and (c) a strong causality by testing a joint F-test of the significance of the short-run coefficients γ_{ij}'s and error correction coefficient δ combined.

The following two points are worth noting in relation to the bounds test. (1) Even though the bounds test approach does not generally require the order of the integration of all the variables to be I(0) or I(1), it does require that none of the variables in the model equation are of order I(2) or higher. In order to ensure that none of the variables are integrated as I(2) or higher, we still need to perform the unit root tests; (2) If some or all of the variables in a regression model are non-stationary in their level form, then the least squares estimation results using a regression model involving these variables may be spurious. As the bounds test results will also reveal whether the variables TA and FDI are cointegrated, we can use the bounds test results to verify whether the model estimation results are spurious or not. However, even if the variables in the regression model are integrated of order 1, that is I(1), if they are cointegrated, then the least squares estimation results would still be valid.

To ensure the three series TA, TR and FDI are either I(0) or I(1) and not I(2) or higher, we perform the unit root (stationarity) test. Table 9.3 presents stationarity test results based on the three commonly used unit root tests, namely the Augmented Dickey-Fuller (1979) test (ADF), Phillips and Perron (1988) unit root test (PP) and KPSS (Kwiatkowski et al., 1992) test results for the three time series variables number of TA, TR and FDI. For the first two tests, the null hypothesis is that there is a unit root in the series under consideration (or the series under consideration is not stationary) and the null hypothesis for the KPSS test is that there is no unit root in the series under consideration (or the series under consideration is stationary).

Table 9.3 Unit root test results

(1)	*ADF (2)*	*p-value (3)*	*Station-ary (4)*	*PP (5)*	*p-value (6)*	*Station-ary (7)*	*KPSS (8)*	*Critical value 5% (9)*	*Station-ary (10)*	*Conclu-sion (11)*
TA	−4.87	0.00	Yes	−4.61	0.00	Yes	0.20	0.15	No	
	−6.76	0.00	Yes	−6.82	0.00	Yes	0.14	0.15	Yes	I(0) or I(1)
TR	0.41	0.99	No	0.41	0.99	No	0.20	0.15	No	
	−3.03	0.04	Yes	−4.75	0.00	Yes	0.10	0.15	Yes	I(1)
FDI	−3.19	0.10	No	−3.57	0.04	Yes	0.20	0.15	No	
	−7.55	0.00	Yes	−4.75	0.00	Yes	0.40	0.46	Yes	I(0) or I(1)
TA	−4.87	0.00	Yes	−4.61	0.00	Yes	0.20	0.15	No	
	−6.76	0.00	Yes	−6.82	0.00	Yes	0.14	0.15	Yes	I(0) or I(1)
TR	0.41	0.99	No	0.41	0.99	No	0.20	0.15	No	
	−3.03	0.04	Yes	−4.75	0.00	Yes	0.10	0.15	Yes	I(1)
FDI	−3.19	0.10	No	−3.57	0.04	Yes	0.20	0.15	No	
	−7.55	0.00	Yes	−4.75	0.00	Yes	0.40	0.46	Yes	I(0) or I(1)

Table 9.4 Bounds test results for testing cointegration

Model (1)	*ARDL Model (2)*	*Null hypothesis (3)*	*F-test Statistic (4)*	*F-test critical values at 5% level of significance*		*Conclusion (7)*
				I(0) (5)	*I(1) (6)*	
F(TA/FDI)	ARDL(1,3)	$H_0: \beta_1 = \beta_2 = 0$ (no cointegration)	11.77	3.94	4.52	TA and FDI are cointegrated
F(FDI/TA)	ARDL(4,5)	$H_0: \beta_1 = \beta_2 = 0$ (no cointegration)	5.68	3.94	4.52	FDI and TA are cointegrated
F(TR/FDI)	ARDL(5,1)	$H_0: \beta_1 = \beta_2 = 0$ (no cointegration)	4.2	3.15	4.11	TR and FDI are cointegrated
F(FDI/TR)	ARDL(1,2)	$H_0: \beta_1 = \beta_2 = 0$ (no cointegration)	148.1	3.94	4.524	FDI and TR are cointegrated

As can be seen, all three series are either stationary in level form or stationery in their first-differences. This means that all three series are either integrated of order 0 or 1, that is the three series are either I(0) or I(1), which is the condition required for use of the ARDL bounds test. This means that the bounds test approach can be used for further investigation.

Note: ADF and PP tests: H_0: series is non-stationary, KPSS: H_0: series is stationary

Table 9.4 presents the bounds test results for cointegration. As can be seen, in all cases, the value of the F-test statistics is larger than the bounds test upper limit – I(1) critical value – at the 5% level of significance. This means that the variables TA and FDI and TR and FDI are cointegrated, respectively.

Table 9.5 presents the corresponding long-run equilibrium relationship and the estimates for the long-run coefficient. The signs of the estimates of the long-run coefficients are positive as they should be and are statistically significant at the 5% level. The estimated coefficient 0.241 for log(FDI) and 1.626 for log(TA) in line 1 of Part A the table means that a 1% increase in FDI will result in an increase of 0.241% in international tourist arrivals and a 1% increase in tourist arrivals will result in an increase of 1.626% in FDI; the estimated coefficient 0.714 and 1.399 in line 1 of Part B of Table 9.5 means that a 1% increase in FDI will result in an increase of 0.714% in international tourism receipts and a 1% increase in international tourism receipts will result in a 1.399% increase in FDI.

Table 9.6 presents the short-run estimation results and the estimates for the long-run error correction terms. As can be seen, all the error correction coefficients are in the range −1 to 0 as they should be and are all statistically significant. This result further confirms the findings of cointegration between TA and FDI and TR and FDI. The first error correction coefficient of –0.188 means that approximately 18.8% of the disequilibrium in the relationship between international TA and FDI caused by previous period shocks are corrected within one period. The other error correction terms can be interpreted in the same way.

Furthermore, the diagnostic test results presented in Table 9.7 reveal that there is no serial correlation or heteroscedasticity issues with our bounds test results. The CUSUM plots presented in Figure 9.3 and the Ramsey RESET test results presented in Table 9.7 reveal that the models are stable.

Table 9.5 Long-run estimation results

Part A: Tourist arrivals				
Dependent variable	$\log(TA_t)$			$\log(FDI_t)$
	Constant	$LOG(FDI_t)$	Constant	$LOG(TA_t)$
Coefficient	3.715	0.241	−2.450	1.626
Standard error	0.457	0.082	1.629	0.321
***p*-value**	0.000	0.006	0.145	0.000
Part B: Tourism receipts				
Dependent variable	$\log(TR_t)$			$\log(FDI_t)$
	Constant	$LOG(FDI_t)$	Constant	$LOG(TR_t)$
Coefficient		0.714	−0.010	1.399
Standard error		0.022	0.184	0.052
***p*-value**		0.000	0.956	0.000

Table 9.6 Short-run estimation results

Dependent variable: Dlog(TA_t)			
ARDL(1,3)	*Estimate*	*Standard error*	*p-value*
DLOG(FDI)	0.066	0.049	0.191
DLOG(FDI(−1))	−0.020	0.017	0.237
DLOG(FDI(−2))	−0.039	0.013	0.004
CointEq(−1)*	−0.188	0.031	0.000
Dependent variable: Dlog(FDI_t)			
ARDL(4,5)	*Estimate*	*Standard error*	*p-value*
DLOG(FDI(−1))	0.301	0.153	0.061
DLOG(FDI(−2))	−0.041	0.169	0.810
DLOG(FDI(−3))	0.179	0.065	0.010
DLOG(TA)	0.237	0.402	0.561
DLOG(TA(−1))	−0.161	0.442	0.718
DLOG(TA(−2))	−0.580	0.420	0.180
DLOG(TA(−3))	−0.697	0.403	0.096
DLOG(TA(−4))	−1.265	0.417	0.005
CointEq(−1)*	−0.378	0.088	0.000
Dependent variable: Dlog(TR_t)			
ARDL(5,1)	*Estimate*	*Standard error*	*p-value*
DLOG(TR(−1))	−0.187	0.145	0.206
DLOG(TR(−2))	0.078	0.140	0.580
DLOG(TR(−3))	0.160	0.124	0.206
DLOG(TR(−4))	0.204	0.125	0.112
DLOG(FDI)	0.303	0.068	0.000
CointEq(−1)*	−0.252	0.086	0.006
Dependent Variable: Dlog(FDI_t)			
ARDL(1,2)	*Estimate*	*Standard error*	*p-value*
DLOG(TR)	1.511	0.270	0.000
DLOG(TR(−1))	0.490	0.261	0.068
CointEq(−1)*	−0.675	0.031	0.000

*1% level of significance

Policy Implications and Concluding Comments

This chapter examined the relationship between FDI and tourism, measured by international tourist arrivals and international tourism receipts, in China. For this purpose, we used annual data over the period 1978–2019 and employed various time series econometric techniques. These include the Granger causality test and the ARDL bounds test, and VECM frameworks. Our results confirm a positive relationship between tourism (measured by

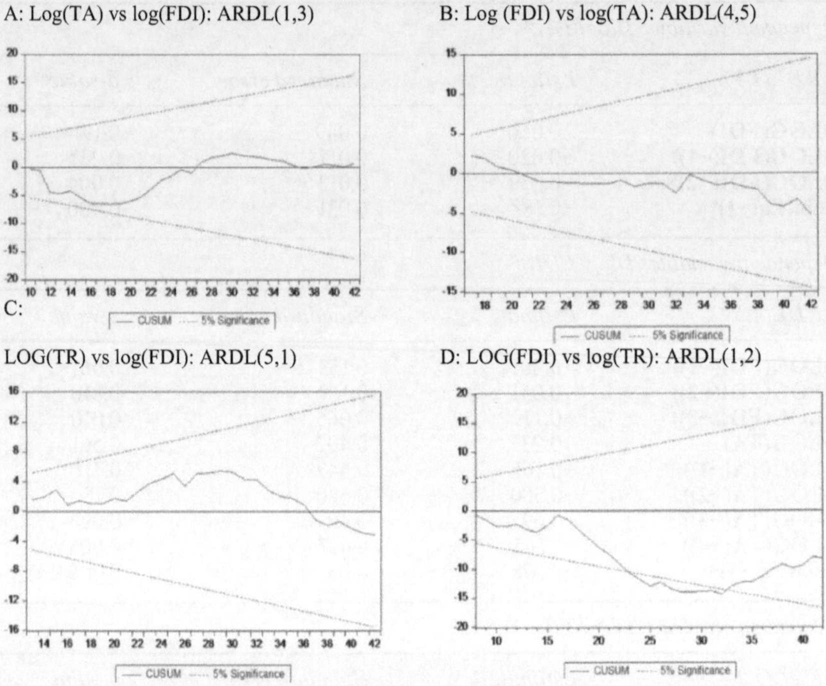

Figure 9.3 CUSUM plots.

Table 9.7 Diagnostic test results

	F-test	*p-value*
Breusch-Godfrey serial correlation LM test		
ARDL(1,3)	0.259	0.774
ARDL(4,5)	1.944	0.165
ARDL(5,1)	0.404	0.672
ARDL(1,2)	1.986	0.121
Breusch-Pagan-Godfrey heteroscedasticity test		
ARDL(1,3)	1.372	0.260
ARDL(4,5)	0.538	0.847
ARDL(5,1)	0.274	0.959
ARDL(1,2)	0.998	0.422
Ramsey RESET test		
ARDL(1,3)	1.172	0.287
ARDL(4,5)	1.272	0.215
ARDL(5,1)	2.281	0.121
ARDL(1,2)	0.079	0.938

international tourist arrivals and tourism receipts) and FDI, as found in most of other studies concerning China and other countries. The Granger causality test results indicated that there is a bidirectional causality between tourist arrivals and FDI as well as tourism receipts and FDI. The long-run estimation results show that a 1% increase in FDI will result in a 0.24% increase in international tourist arrivals and a 1% increase in international tourist arrivals will result in a 1.63% increase in FDI; a 1% increase in FDI contributes a 0.71% increase in tourism receipts and a 1% increase in tourism receipts leads to a 1.40% increase in FDI.

These results provide some useful insights for the tourism and FDI promotion policies in China. In particular, these results confirm that FDI plays a vital role in stimulating tourism growth, while tourism also presents as an effective tool to attract FDI into China. Enhanced tourism as well as FDI inflows would undoubtedly contribute to economic growth in China. Therefore, further emphasis in providing a conducive environment to attract FDI and strategies to attract more international tourists are encouraged.

Nevertheless, FDI and tourism growth are not without concerns. When considering FDI and environmental pollution, literature shows mixed results; some studies have shown that FDI inflows are positively related with energy use and CO_2 emissions (Nadeem et al., 2020; Shahbaz et al., 2018), while most of the research on China suggests that there is no significant positive relationship between FDI and environmental pollution and, in fact, there is an increase in the air quality in some instances (Liu et al., 2018; Kirkulak et al., 2011; Jiang et al., 2018). However, Liu et al. (2018) do note that FDI inflows had distinct effects on different environmental pollutants. In particular, FDI inflows were associated with reduced waste soot and dust pollution, while it increased the degree of wastewater and sulfur dioxide pollution. To this end, promoting further FDI inflows to China will not only contribute to the development of the tourism industry but have limited or no negative implications on environmental quality. Therefore, it may be worthwhile to provide incentives to foreign investors in tourism and other sectors with stringent environmental rules to redirect the inflows of FDI into dirty industries in China (Jun et al., 2018).

There is a growing body of literature that highlights the environmental implications of tourism development. For example, a significant body of literature that confirms the causal relationship running from tourism to CO_2 emissions world-wide and hence, the tourism industry is regarded as a major contributor to climate change (see, for example, Durbarry & Seetanah, 2014; Katircioglu, 2014; Selvanathan et al., 2020; Gossling et al., 2013). However, the counterargument for this idea is that in a situation of considerably higher levels of income growth, followed by high levels of FDI growth and the existence of a high performing tourism industry, for example, the level of CO_2 emissions may be mitigated, supporting the Environmental Kuznets Curve Hypothesis

(EKCH) (Grossman & Krueger, 1991; De Vita et al., 2015). However, some research on China confirms the former, where tourism contributes to CO_2 emissions and environmental pollution (Tang et al., 2014; Sharif et al., 2020).

Given the pace of expansion of the tourism sector in China and its potential impact on the environment, there is an increasing responsibility for tourism-related stakeholders to manage tourism in a sustainable manner. In this regard, Scott et al. (2016) suggested that a strategic tourism development policy that explicitly promotes sustainable tourism and low carbon emissions, with specific targets to be achieved within a specific time frame is vital. To assist tourism industry stakeholders to achieve these targets on time, Scott et al. (2016) suggest that governments should facilitate the achievement of targets by means of establishing emissions monitoring mechanisms through a clear emissions measurement and reporting system, providing financial and technical support to achieve low-carbon tourism promotion and management. Another proposed strategy is the active promotion of low-carbon tourism demand where tourists will play a significant role in helping carbon emissions reduction through low-carbon sustainable consumption patterns with respect to transportation, sightseeing, accommodation, shopping and entertainment (Tang et al., 2011). This is particularly important for a country such as China with more than 140 million international tourist arrivals per year.

The energy-intensive nature of the tourism industry is the main reason for high level of tourism-related CO_2 emissions. Therefore, necessary incentives should be extended to all tourism stakeholders to adopt alternative energy sources that are environmentally friendly. In particular, greater use of cleaner energy transport solutions should be promoted in China as long-distance road transport is a common feature in its tourism industry. Furthermore, Tang et al. (2018) found that tourism transport had the greatest impact on the tourism industry-related CO_2 emissions in China. The timely attention of policy makers towards the issues highlighted in this study and recommended actions will promote sustainable tourism and thereby attract high levels of FDI and promote sustainable development in China.

Acknowledgement

The authors wish to acknowledge the internal grants from Griffith Asia Institute, Griffith University, Australia and Asia Pacific College of Business and Law, Charles Darwin University, Australia to undertake this research project. The authors would like to thank Ms. Tanya Parker for her editorial assistance.

References

Amin, S., Kabir, F. A., Nihad, A. & Khan, F. 2020. An Empirical Investigation between Foreign Direct Investment (FDI) and Tourism in Bangladesh. *Journal of Tourism Management Research,* 7, 110–121.

Arain, H., Han, L., Sharif, A. & Meo, M. S. 2019. Investigating the Effect of Inbound Tourism on FDI: The Importance of Quantile Estimations. *Tourism Economics,* 26, 682–703.

Bezić, H. & Nikšić Radić, M. 2017. Tourism Foreign Direct Investment Led Tourism Gross Value Added: A Cointegration and Causality Analysis of Croatian Tourism. *Economic Research-Ekonomska Istraživanja,* 30, 1443–1460.

Chen, X. 2010. The Influence of FDI on China's Tourism Industry. Auckland University of Technology, Auckland.

Chen, Y. 2017. China's Tourism-Led Foreign Direct Investment Inflows: An Empirical Study. *Modern Economy,* 8, 39–50.

Chinese Statistical Yearbook. 2019. China Statistical Press. Retrieved from http://www.stats.gov.cn/tjsj/ndsj/2019/indexeh.htm.

Choi, H.-C. & Sirakaya, E. 2006. Sustainability Indicators for Managing Community Tourism. *Tourism Management,* 27, 1274–1289.

Das, D. K. 2007. Foreign Direct Investment in China: Its Impact on the Neighboring Asian Economies. *Asian Business & Management,* 6, 285–301.

De Vita, G., Katircioglu, S., Altinay, L., Fethi, S. & Mercan, M. 2015. Revisiting the Environmental Kuznets Curve Hypothesis in a Tourism Development Context. *Environmental Science and Pollution Research,* 22, 16652–16663.

Dickey, D. & Fuller, W. 1979. Distribution of the Estimators for Autoregressive Time Series with a Unit Root. *Journal of the American Statistical Association,* 74, 427–431.

Dunning, J. H. & Mcqueen, M. 1981. The Eclectic Theory of International Production: A Case Study of the International Hotel Industry. *Managerial and Decision Economics,* 2, 197–210.

Durbarry, R. & Seetanah, B. 2014. The Impact of Long Haul Destinations on Carbon Emissions: The Case of Mauritius. *Journal of Hospitality Marketing & Management,* 24, 401–410.

Dwyer, L. & Forsyth, P. 2008. Economic Measures of Tourism Yield: What Markets to Target? *International Journal of Tourism Research,* 10, 155–168.

Garcia-Flores, F., Pe'Rez-Cirera, V., Martı'N, P., Alanis, G. & Guadarrama, L. 2008. Tourism, Trade and the Environment: Tourism and Coastal Development in the Mexican Portion of the Mesoamerican Reef. *Fourth North American Symposium on Assessing the Environmental Effects of Trade, Commission for Environmental Cooperation. Montreal.*

Gossling, S., Scott, D. & Hall, C. M. 2013. Challenges of Tourism in a Low-Carbon Economy. *WIREs Climate Change,* 4, 525–538.

Granger, C. W. J. 1986. Developments in the Study of Cointegrated Economic Variables. *Oxford Bulletin of Economics and Statistics,* 48, 213–228.

Grossman, G. M. & Krueger, A. B. 1991. Environmental Impacts of a North American Free Trade Agreement. *National Bureau of Economic Research,* No. w3914.

Gujarati, D.N. 1995 Econometrics. 3rd Edition, McGraw-Hill, Inc., New York.

Gupta, V. K. 1983. A Simultaneous Determination of Structure, Conduct and Performance in Canadian Manufacturing. *Oxford Economic Papers,* 35, 281–301.

Jayasinghe, M. & Selvanathan, E. A. 2021. Energy Consumption, Tourism, Economic Growth and CO_2 Emissions Nexus in India, *Journal of the Asia Pacific Economy, 26*(2), 361–380.

Jiang, L., Zhou, H.-F., Bai, L. & Zhou, P. 2018. Does Foreign Direct Investment Drive Environmental Degradation in China? An Empirical Study based on Air Quality Index from a Spatial Perspective. *Journal of Cleaner Production,* 176, 864–872.

Jun, W., Zakaria, M., Shahzad, S. J. H. & Mahmood, H. 2018. Effect of FDI on Pollution in China: New Insights Based on Wavelet Approach. *Sustainability,* 10, 3859.

Katircioglu, S. 2011. The Bounds Test to the Level Relationship and Causality between Foreign Direct Investment and International Tourism: The Case of Turkey. *E + M Ekonomie a Management*, 14, 6–13.

Katircioglu, S. T. 2014. International Tourism, Energy Consumption, and Environmental Pollution: The Case of Turkey. *Renewable and Sustainable Energy Reviews,* 36, 180–187.

Kirkulak, B., Qiu, B. & Yin, W. 2011. The Impact of FDI on Air Quality: Evidence from China. *Journal of Chinese Economic and Foreign Trade Studies,* 4, 81–98.

Kundu, S. K. & Contractor, F. J. 1999. Country Location Choices of Service Multinationals: An Empirical Study of the International Hotel Sector. *Journal of International Management,* 5, 299–317.

Kurtenbach, E. & Oslon, A. 2020. China virus outbreak rams global tourism, costing billions. AP News [Online]. Available: https://apnews.com/article/70ded34c495a87e4f0e006fd52e64e6b [Accessed 11th December 2020].

Kwiatkowski, D., Phillips, P. C. B., Schmidt, P. & Shin, Y. 1992. Testing the Null Hypothesis of Stationarity Against the Alternative of a Unit Root. *Journal of Econometrics,* 54, 159–178.

Lim, C. 1997. An Econometric Classification and Review of International Tourism Demand Models. *Tourism Economics,* 3, 69–81.

Liu, Q., Wang, S., Zhang, W., Zhan, D. & Li, J. 2018. Does Foreign Direct Investment Affect Environmental Pollution in China's Cities? A Spatial Econometric Perspective. *Science of the Total Environment,* 613–614, 521–529.

Nadeem, A. M., Ali, T., Khan, M. T. I. & Guo, Z. 2020. Relationship between Inward FDI and Environmental Degradation for Pakistan: An Exploration of Pollution Haven Hypothesis through ARDL Approach. *Environmental Science and Pollution Research,* 27, 15407–15425.

Pesaran, M. H., Smith, R. J. & Shin, Y. 2001. Bounds Testing Approaches to the Analysis of Level Relationships. *Journal of Applied Econometrics,* 16, 289–326.

Phillips, P. & Perron, P. 1988. Testing for a Unit Root in Time Series Regression. *Bimetrika,* 75, 335–346.

Samimi, A. J., Sadeghi, S. & Sadeghi, S. 2017. The Relationship between Foreign Direct Investment and Tourism Development: Evidence from Developing Countries. *Institutions and Economies,* 5(2), 59–68.

Sanford, D. M. & Dong, H. 2000. Investment in Familiar Territory: Tourism and New Foreign Direct Investment. *Tourism Economics,* 6, 205–219.

Scott, D., Gössling, S., Hall, & Peeters, P. (2016). Can Tourism be Part of the Decarbonized Global Economy? The Costs and Risks of Alternate Carbon Reduction Policy Pathways. *Journal of Sustainable Tourism,* 24(1), 52–72.

Selvanathan, E. A., Jayasinghe, M. & Selvanathan, S. 2020. Dynamic Modelling of Inter-relationship between Tourism, Energy Consumption, CO_2 Emissions and Economic Growth in South Asia. *International Journal of Tourism Research,* 22(4), 597–610.

Selvanathan, S., Selvanathan, E. A. & Viswanathan, B. 2012. Causality Between Foreign Direct Investment and Tourism: Empirical Evidence from India. *Tourism Analysis,* 17, 91–98.

Shahbaz, M., Nasir, M. A. & Roubaud, D. 2018. Environmental Degradation in France: The Effects of FDI, Financial Development, and Energy Innovations. *Energy Economics,* 74, 843–857.

Shan, J. & Wilson, K. 2001. Causality between Trade and Tourism: Empirical Evidence from China. *Applied Economics Letters,* 8, 279–283.

Sharif, A., Saha, S., Campbell, N., Sinha, A. & Ibrahiem, D. M. 2020. Tourism, Environment and Energy: An Analysis for China. *Current Issues in Tourism,* 23, 2930–2949.

Song, C., Shi, S., Chen, J. M. & Li, X. 2019. Does Inward FDI Affect Outward FDI in Tourism in China? *Tourism Economics,* 26, 344–349.

Tang, S., Selvanathan, E. A. & Selvanathan, S. 2007. The Relationship between Foreign Direct Investment and Tourism: Empirical Evidence from China. *Tourism Economics,* 13, 25–39.

Tang, Z., Bai, S., Shi, C., Liu, L. & Li, X. 2018. Tourism-Related CO_2 Emission and Its Decoupling Effects in China: A Spatiotemporal Perspective. *Advances in Meteorology,* 2018, 1473184.

Tang, Z., Shang, J., Shi, C., Liu, Z. & Bi, K. 2014. Decoupling Indicators of CO_2 Emissions from the Tourism Industry in China: 1990–2012. *Ecological Indicators,* 46, 390–397.

Tang, Z., Shi, C.B., & Liu, Z. 2011. Sustainable Development of Tourism Industry in China Under the Low-carbon Economy, *Energy Procedia*, 5, 1303–1307.

Tisdell, C. & Wen, J. 1991. Investment in China's Tourism Industry: Its Scale, Nature, and Policy Issues. *China Economic Review,* 2, 175–193.

Tsang, N. K. F. & Hsu, C. H. C. 2011. Thirty Years of Research on Tourism and Hospitality Management in China: A Review and Analysis of Journal Publications. *International Journal of Hospitality Management,* 30, 886–896.

Witt, S. F. & Witt, C. A. 1995. Forecasting Tourism Demand: A Review of Empirical Research. *International Journal of Forecasting,* 11, 447–475.

Zapata, H. O. & Rambaldi, A. N. 1997. Monte Carlo Evidence on Cointegration and Causation. *Oxford Bulletin of Economics and Statistics,* 59, 285–298.

10 Foreign Direct Investment in Nature Tourism

A Proposed Model for Economic Development of Bangladesh

Sara Tasneem

Introduction

Foreign direct investment endows the prospects to play a significant role to accelerate the growth and development of tourism industry, specifically nature tourism. Foreign Direct Investment (FDI) can uncover the potential of the tourism sector of Bangladesh. Tourism growth is predicted to rise with the utilisation of FDI in the near future, akin to the manufacturing sector. Gradually, tourism has become a contributing sector to Bangladesh's economy (WTTC, 2017). Increasing contribution of tourism to domestic production indicates the vital importance of the sector to economic development which determines the economic well-being of a nation. FDI has prospects to extract the potential of nature tourism, leading to economic development. In a labor intensive country like Bangladesh, FDI can be a medium of exchange of human resource generation, capital augmentation, utilisation of scarce natural resources, more integration of the domestic economy with the overseas economy, and achievement of macroeconomic goals. FDI can play a significant role in spurring the growth and development of newly emerging resource sectors of the country along with other growth. FDI can assist in accumulating national savings and investment through the creation of a competitive investment environment in the host country. FDI became important in world economic development from the mid of 1980s (Baker, 1998). Bangladesh has received a thin flow of FDI in the 1970s and 1980s. The scenario began to change since the early 1990s. Being privileged in geographical location, low labor cost, low level of initial investment in fixed and variable assets, greater domestic market, and flexible policy and investment incentives, Bangladesh invites FDI in the manufacturing sector. Consequently, the FDI inflow has noteworthy contribution for industrial sector growth and development. Even though industrial sector growth is considered a pre-requisite of economic growth, Bangladesh has other sectors which are yet to be developed for economic growth. Bangladesh economy can create employment opportunity, get tax revenue, increase foreign currency reserve, build up the infrastructure, develop the remote region, find new areas of trade, and swap cultural and social values with other nations

DOI: 10.4324/9781003155492-13

through FDI in nature tourism. In spite of being one of the promising industries of Bangladesh, tourism is still developing. Being a universal economic activity, it recurrently contributes to the domestic and global economy and the industry is growing rapidly in Bangladesh (Blanke & Chiesa, 2007). As Bangladesh is well-known for its natural beauty (Ali, 2004), it has high potential in the Asian region for tourism development leading to national output growth (Islam & Islam, 2004).

The share of tourism to the Gross Domestic Product (GDP) of Bangladesh in 2019 was 2.21% (WTTC, 2019). Out of total employment, Bangladesh's tourism absorbs only 1.91% of total labor force (WTTC, 2019). In 2020, the sector occupies 1.4% of total investment. During the last decades, the government has increased spending on the sector at a negligible rate, from 2.03% in 2001 to 2.23% in 2020 out of total spending (WTTC, 2019). The poor statistics represent the scenario of tourism in Bangladesh. The sector needs noteworthy investment to augment its growth and contribution to economy. The sluggish domestic investment and government expenditure induce the demand of FDI in tourism. In Bangladesh, the scope of nature tourism, culture tourism, and eco-tourism is quite evident (Shamsuddoha, 2004).In this connection, nature tourism will be a profitable area to ensure augmentation of the entire area. FDI in nature tourism is expected to promote regional development leading to national economic growth. Economic growth determines the well-being of a nation. From the ancient era to present, nature tourism in Bangladesh has an appeal to domestic and foreign tourists. Overseas investors can utilise the potential of nature tourism compared to other sectors of the economy at a lower cost and effort as it directly belongs to natural resources which need not be processed further. In the light of the aforesaid discussion, the study aims at exploring the potential of nature tourism and developing a model to show how the utilisation of FDI in natural sites promotes nature tourism in order to achieve economic development. The remaining sections portray the literature review, research questions, statement of objectives, methodology, potential of nature tourism in Bangladesh, pattern of FDI, and the model.

Literature Review

The present section has conducted an in-depth review of existing empirical studies. A critical review on studies related to tourism industry and the role of FDI in Bangladesh's economy has been conducted. In the end, a research question has been floated regarding the utilisation of FDI in natural sites to promote nature tourism leading to national economic development.

Sultana (2016) explores the contribution of tourism industry to Bangladesh's economy. Tourism is a growing and key sector for economic development. The industry tries to achieve the macroeconomic goals, i.e., generation of income, creation of employment opportunity in hotels, restaurant, leisure industries of transportation, receiving foreign exchange, and augmenting public and private spending. The study observes the need of capital investment for the rapid growth of the industry. Finally, it ends with the recommendation of infrastructure and super infrastructure building developing entrepreneurial activities, increasing market share, augmenting investment, and implementation of training programs.

Alauddin et al. (2014) examine the prospects of the tourism industry of Bangladesh. Tourism plays a significant role in economy. The natural beauty, archeological sites, historical heritage, hospitality of citizens, and low cost of labor are the strengths of the industry. Despite these strengths, the sector is passing through obstacles such as lack of new domestic investment, absence of professional human resource, poor country image, lack of safety measures, FDI unavailability, and improper promotional activities. The sector is exposed to the threats of political fluctuation, tourism policy, natural disasters, lack of communication skill of local citizens, and infrastructural bottlenecks. Research activity, long term policy, information availability, and global connectivity are the opportunities to develop the industry. The study recommends sufficient promotional support, attracting FDI and domestic private investment, coordination among concerned entities, exemption of purview tax and VAT, government initiative for infrastructural development, simplification of immigration formalities, building training centers, and fund allocation in national budget. It concludes uncovering the immense significance of tourism development from an economic, social, political, and cultural perspective.

Jahan & Amin (2014) analyse the factors related to sustainability of tourism development in Bangladesh. Sustainability refers to dimensions such as the environmental, socio-cultural, and economic aspects of tourism development. A proper balance must be maintained between the three dimensions to ensure sustainability in the long run. Sustainability aims at minimising tourists' impact on the places visited, encouraging the safeguard of domestic heritage, and protecting the natural environment. Integrated participation of all stakeholders with sound political leadership is required for sustainability. The study selects one of the prosperous areas for nature tourism in Bangladesh, namely Sylhet. It aims to find out tourists' experience while visiting Sylhet, identification of positive and negative issues affecting tourists, assessment of tourists' sensitivity to environmental pollution, and recommendations to promote tourism sustainability in the study area. The study utilises primary and secondary sources to collect data. Allowing a descriptive nature, it finalises 26 variables to prepare the survey questionnaire. The five point Likert scale was used to assess the importance of the items, following the non-probability judgment sampling technique. The Cronbach's Alpha score of 0.777 indicates the reliability of 329 responses. An index value of 0.6007 (25 variables) reveals that Sylhet has the potential for sustainable tourism development. The results reveal the factors such as purchasing preference, supporting services, service quality, impact of sustainable tourism, involvement of concerned authority, responsibility, resource and environment, and knowledge and awareness of tourists have notable influence on the future of Sylhet to be a sustainable tourism destination. The study concludes with the recommendation of proper maintenance of the natural and cultural heritage of Sylhet, strict monitoring of utilisation of resources and consideration of tourists' tastes, preference, and fashion to ensure sustainability in the long run.

The tourism industry makes significant contributions to the economy of a developing country like Bangladesh. Tourism sector development promotes economic growth, utilises natural resources, minimises unemployment rate, augments local and regional growth, reduces inequality, expands trade and commerce, earns foreign currency, and improves the standard of living of citizens. Hafsa (2020) examines tourism growth and the economic contribution of the industry to Bangladesh's economy. The study observes growth in tourism in recent decades and the erratic pattern of growth implies lack of proper planning at national level. Bangladesh has low employment absorption in tourism industry in comparison to other Asian countries. The study reveals the strengths, weakness, opportunities, and threats to the sector. Existence of natural resources is a notable strength. Whereas poor fund and promotional activities, lack of proper tourism policy, insufficient safety and security are considered significant problems for the sector. The study concludes with the recommendation of infrastructural development.

Development of tourism industry is a multi-dimensional issue which impacts the entire economy. It requires sound infrastructure, smooth communication systems, local entrepreneurial growth, support of utilities, transport availability, and flexible procedure both for domestic and international tourists. Hasan &Hossain (2014) opine that the tourism industry can be developed through the growth of Small and Medium Enterprises (SME). The eight divisions of Bangladesh provide a versatile natural destination for tourists. Local SMEs linked to tourism face a number of barriers. SMEs are contributing to other sectors but in tourism, the contribution is lower than expected. So, the study explores the potential, problem, and present scenario of tourism SME. Primary data have been collected through a structured questionnaire. The results reveal that tourism SMEs bear a more prominent role in economic development. Tourism is often a heritage of local area that exists in micro-destination. Out of the total, 90.4% male owners and 9.6% female owners are engaged in SME tourism. The owners of tourism SMEs have deficiency of IT skills. The initial capital comes from personal and family sources and owners face difficulties in accessing financial institutions. The provision of training for owners regarding entrepreneurial activities is poor. The study emphasises well-developed tourism SMEs for the rapid development of Bangladesh's tourism industry.

Hossain (2021) analyses the influence of macroeconomic determinants on the inflow of FDI. FDI is desired for a country suffering from capital poverty. For an emerging country like Bangladesh, FDI is a trajectory to facilitate knowledge accumulation and industrial development. The determinants of FDI inflow are dynamic and change overtime for the recipient country. The study employs data from 1975 to 2015 to estimate the autoregressive distributed lag model and judge the short and long run relationship between foreign direct investment, current account balance, export, import, per capita gross domestic product, and total foreign currency reserves. The results reveal a long run relationship between FDI and macroeconomic determinants. The development in communication may have positive correlation with FDI. The results also indicate high production cost in electricity and inconsistence production level due to dependence on external factors. The study recommends the use of renewable energy and adoption of FDI-led development strategy.

Mondal (2003) analyses the importance of FDI, perceptions of investors, and the role of government to attract FDI in Bangladesh. The author opines that FDI positively influences the short run and long run development of global business. Bangladesh has the opportunity to reap the benefits of FDI by advancing the investment scenario of the country. The presence of non-performing loans and credit deficiency plague the financial system of the country. Additionally, the loss incurred by state owned enterprises has a detrimental effect on FDI inflow. The study observes lack of government initiatives to attract substantial FDI in 1990s. In order to understand the perceptions of entrepreneurs, a questionnaire has been designed and a survey undertaken in Dhaka, Sylhet, and Chittagong. The survey was targeted to extract the opinion on the political, financial, and regulatory status of Bangladesh to attract FDI. Applying the discriminate analysis on collected data, the study reveals the perceptions of the entrepreneurs. The political factors rank the most important one to limiting FDI. Privatisation pace is expected to be faster to attract FDI. To limit FDI, the 'cost of doing business' is a significant factor. The study recommends improvement of capital market, acceleration of privatisation, and prudent examination of the cost factor related to FDI.

Rahman (2015) evaluates the impact of FDI on the economic development of Bangladesh. FDI can fill the capital gap in a developing country like Bangladesh. The FDI inflow has been significantly increasing since 1999 with the new industrial policy taken by Bangladesh's government. FDI has been encouraged in manufacturing sectors except the reserved industries. Bangladesh offers non-discriminatory benefits to local and foreign entrepreneurs. The overseas investors enjoys tax holiday, payment of royalties, 100% equity and profits. The government has established EPZs to influence the FDI inflow. Despite the incentives, there are factors which have a detrimental impact on FDI. So, the study aims at finding the negative factors of FDI and impacts of FDI on economic development. The multiple regression analysis has been conducted using the data from the years 1999–2013. The results indicate an insignificant positive relationship between FDI and economic growth. There is significant positive and negative correlation between FDI and inflation rate and balance of trade, respectively.

FDI is considered to be a prospective weapon of economic growth. It assists in accumulating capital, transmitting technology, and broadcasting production capacity and the interaction of domestic and global economies. Dey and Awal (2017) investigate the relation between FDI and economic growth. For the investigation, time series data for the years 1990–2015 have been used. The multiple regression equation has been estimated using the ordinary least squares technique. The diagnostic tests for normality, heteroskedasticity, autocorrelation, and residual have been conducted. The analysis investigates the impacts of FDI, remittance, import, export, and inflation on GDP growth. The results of the diagnostic tests reveal that the regression model is satisfactory as 96% variation in GDP can be explained by the selected regressions. The data follows a normal distribution and confirms homoskedasticity and zero serial correlation. The study finds a negative relation between FDI and growth. The negative relation may result from insufficient inflow and low level of skilled human resources. Nonetheless, political environment was detrimental to FDI inflow. The study suggests political stability, utility supply, human resource development, and good governance to attract FDI.

Monzoor and Chowdhury (2017) scrutinise the impact of FDI on some macroeconomic indicators and examine the relation between FDI and the indicators. FDI is viewed as potent driver for economic development. It is a prime source of capital flow and knowledge dissemination. To fill the gap of domestic capital, Bangladesh has encouraged FDI in private sector from the 1980s. The objectives behind FDI liberalisation are augmenting growth, employment expansion, and generating new sources of capital. The study investigates the effect of FDI on GDP, imports, exports, and total formation of capital. The data for the years 1994–2014 have been analysed. The findings suggest no unique relation between FDI and growth. FDI has a positive impact on growth and capital at an earlier stage followed by a declining phase. Further, in the later period, the relation becomes positive.

The above discussion ends with a research question on the utilisation of FDI generally on tourism sector and particularly in nature tourism leading to national economic development. The study has defined nature tourism in a differentiated manner while exploring the potential of nature tourism. The new way of viewing nature tourism is a unique contribution of this study. Though notable studies have been carried out, there is no specific attempt to demonstrate a conceptual model (following the inductive reasoning approach) portraying probable positive impacts of FDI on tourism in order to achieve economic development. Hence, an attempt has been made to fill the gap.

Research Questions

The study has the following research questions:

- What is the potential of nature tourism in Bangladesh?
- How can FDI promote economic development through progress in nature tourism?

Objectives

The study has the following two objectives:

- To explore the potential of nature tourism
- To develop a model to show how the utilisation of FDI in natural sites promotes nature tourism in order to achieve economic development

Methodology

The approach to the investigation of the research questions is inductive reasoning with the observation of tourism industry and pattern of FDI in Bangladesh. The exploration of tourism industry gives birth to the identification of thrust areas in tourism and the necessity of FDI. In-depth exploration found nature tourism lacking in investment. The study aims at exploring the potential of nature tourism. A model has been designed to depict the positive impacts of FDI on nature tourism and the probable economic development through initiating progress in nature tourism. Further exploration of natural sites of beauty finds out the places which have potential for nature tourism development. The data collected from secondary sources help to observe the pattern of FDI, FDI inflow, composition of FDI, domestic sectors receiving FDI, and top foreign investors in Bangladesh. To date, any theoretical and methodological basis is nonexistent to the emerging issue of nature tourism for Bangladesh. Hence, reviewing empirical literatures is a part of methodology. To find the solution for the research questions, it has arrived at a general explanation – a conceptual model. In order to attain the objectives, potential of natural sites has been evaluated and a model has been proposed. The study exemplifies applied research and the analysis is mostly qualitative in nature.

Potential of Nature Tourism in Bangladesh

Tourism is the commercial arrangement and activities of holidays to visits the places of curiosity and interest in order to attain pleasure. Bangladesh has the potential to develop tourism industry. Since the ancient period, domestic and international tourists have been traveling the country to enjoy the enormous natural beauty. Bangladesh is an attractive destination for tourists and suitable platform for tourism industry as it boasts natural beauty, archeological sites, historical places, the longest beach, the world's largest mangrove forest, and other man-made structures, cultural practices, social values, besides a variety of lucrative products. The elements of natural beauty have uncovered a new horizon for the industry. So the study defines nature tourism in a slightly differentiated manner. Nature tourism can be defined as the arrangement of accommodation, infrastructure, and communication in each natural site of the country for tourists' destination with environmental safety in mind. It will promote local economic progress at district level and augment the development of the entire tourism industry. Bangladesh is endowed with an array of natural sites such as lakes, canals, *haor-baor*, rivers, the Bay of Bengal, tropical rain forest, the mangrove forest, hills and mountains, and pleasant islands. The aforementioned elements are scattered throughout the country. So, the preparation for natural site development will automatically embrace expansion of local trade and socio-economic development. In addition, nature tourism allows for the decentralisation of the industry which will assist to reap the economic benefits for all classes of citizens of the country. In order to uncover the potentials of nature tourism, the study has categorised natural beauty in the following manner (Table 10.1).

The inland water resources consists of the main rivers (namely the Padma, the Meghna, and the Jamuna), their tributaries, the *haors* (the Saneer *haor*, Hakaluki *haor*, Tanguar *haor*, etc.), the *beels* (Arial *beel*, Chalan *beel*, Katla

Table 10.1 Natural sites of Bangladesh

Natural beauty				
Natural sites	The inland and marine water resources	The tropical forests and the mangrove forests	The hills and mountains	The islands
Tourism category	River and marine tourism	Forest tourism	Mountain tourism	Island tourism
Activities (related to natural sites) to be accomplished	Resort building, transportation service, infrastructure building, utility service, service of hygienic food, arrangement of regular fair of travelers' products, initiatives to protect the tourists, services to preserve natural beauty			
The districts having natural sites	Rajshahi, Pabna, Narayangang, Tangail, Chandpur, Faridpur, Brahmanbaria, Mymensingh, Gazipur, Sylhet, Sunamganj, Hobigong, Moulovibazar, Khulna, Satkhira, Bagherhat, Potuakhali, Cox's Bazar, Chittagong, Comilla, Rangamati, Khagrachhari, Bandarban, Bhola, Netrakona, Sherpur, Rangpur, Munshiganj, Chattogram, Kishoreganj.			

Source: Author's data.

beel, Nagarkanda *beel*, etc.), and the waterfalls (Madhab Kunda, Jadipai, Hum Hum, Khoiyachara, Amiakhum, etc.). All the districts are feasible platforms for river tourism in Bangladesh. The Bay of Bengal to the south of the country is the center of marine tourism. Besides the mangrove forest 'Sundarban', Bangladesh hosts the hill forests, mangrove plantations, and *sal* forest. Bangladesh is endowed with approximately 75 mountains located mainly in the south-eastern part of the country. The hilly districts of the south-eastern part the country, Rangamati, Khagrachhari, and Bandarban are examples. The islands and char areas are scattered along the Bay of Bengal and the river Padma. The islands, Nijhum Dwip, Bhola, Sandwip, St. Martin's, Kutubdia, Maheskhali, Sonadia, Hatiya, and Chhera Dwip can be attractive to tourists as if developed well. The southern, south-eastern, north-eastern, and north-western part of the country allow for nature tourism development.

Pattern of FDI in Bangladesh

FDI is a special category of investment which controls the ownership of business within the boundary of one country by an entity located in another country. The firms of domestic economy can enter the arena of international business to introduce investment in any overseas economy. It is a bridge of knowledge and technology sharing which assists to spur the economic growth of the host country. It takes the form of equity capital, investment of retained earnings, and commercial loans to the firms of domestic economy. Bangladesh has been trying to receive FDI since the 1980s. Immediately after independence, Bangladesh receives almost no FDI due to its nationalisation policy. Post the new industrial policy of the 1980s, Bangladesh had started to receive FDI. In the 2000s and 2010s, the FDI inflows were increasing (Figure 10.1).

Bangladesh's government promotes FDI to expand the growth of the private sector. The equity capital, reinvested earnings, and intra-company

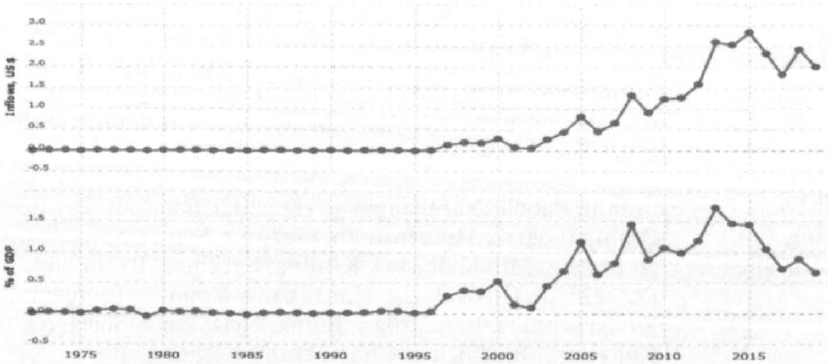

Figure 10.1 FDI Inflows from 1972 to 2019.

Table 10.2 Status of FDI inflow and greenfield investment in 2019

FDI	*Year 2019*
FDI inward flow (million USD)	1,597
FDI stock (million USD)	16,385
Number of greenfield investments	18
Value of greenfield investment (million USD)	5,762

Source: UNCTAD, 2020.

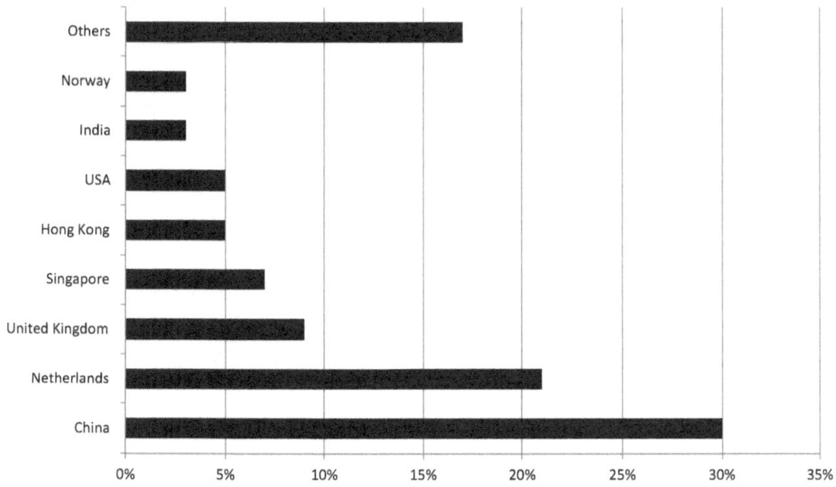

Figure 10.2 Top investor countries in 2019 (July–June).

loans of FDI in 2019 were US$394.73 million, US$796.01 million, and US$ −8.45 million respectively (Bangladesh Bank, 2019). The country allows greenfield investment where a foreign parent company can start a new venture by constructing new functional facilities in Bangladesh (Table10.2).

Bangladesh is becoming a promising country for foreign investors. Overseas countries are recognising Bangladesh as an emerging market for economic growth, strategic geographical location, low paid skilled labor force, trade openness, and increasing speed of digitalisation. The number of investor countries is increasing gradually. Singapore, China, Hong Kong, USA, Japan, and Malaysia are the countries which invest a substantial amount. China was the largest source of FDI in the fiscal year 2019 (Figures 10.2 and 10.3).

The Model

The proposed conceptual model is a representation of concepts and system which is used to understand the utilisation of FDI in natural sites to promote nature tourism in order to achieve economic development. It includes

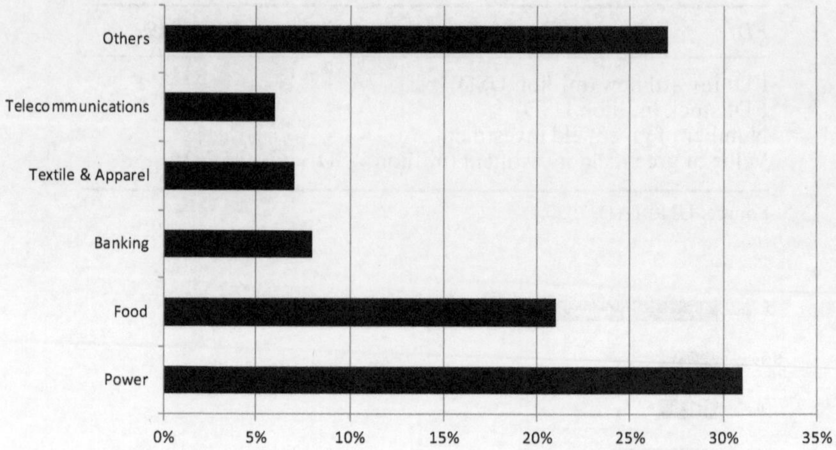

Figure 10.3 Top sectors using FDI in Bangladesh during 2019 (July–June).

the theme, assumptions, illustration, diagrammatic presentation, arguments in favor of the model, challenges, and finally the recommendations to overcome the challenges. It is named PFT model (Prospects of FDI in Nature Tourism).

Theme

Nature tourism is a new horizon to attract and utilise FDI. The utilisation of FDI in nature tourism promotes development of the tourism industry and Bangladesh's economy. Hence, in the long run, Bangladesh will be a potential tourist destination. Moreover, prospects of FDI in nature tourism will augment FDI inflows from new foreign investors implying accelerated economic development.

Assumptions

The model requires the following assumptions:

- Bangladesh's government is continuously working to attract foreign investors. Hence, the obstacles to FDI are regularly alleviated
- Good coordination exists among the state-owned enterprises
- Public utilities (water, gas, and electricity) are smooth and continuous
- Local citizens maintain brotherhood, harmony, and courtesy with overseas citizens
- Domestic entrepreneurs are encouraged to invest in the tourism industry along with foreign entrepreneurs

Illustration

The model has been divided into three phases. In the first phase a new arrangement of investment is required. Domestic investors are experienced to take the responsibility of manufacturing and service sector. Moreover, they feel less worried to tackle the problems of the domestic economy. The remaining sector, agriculture, is a place for the entire population to provide their talents and efforts. The existing foreign investors are encouraged to continue business operation and allocate new funds to invest in national infrastructure. National smooth infrastructure is the way to reach the natural sites. So, infrastructural development must be ensured. Finally, the economy will attract new foreign investors to invest in natural sites for tourism development. The second phase portrays the procedures which illustrates how FDI will assist to promote nature tourism. FDI in the form of greenfield investment is designed to motivate foreign investors, whereas commercial loans protect the interests of both local and foreign entrepreneurs. Phase three reveals the results of the model. If proper implementation is ensured, Bangladesh will be a new destination owing to nature tourism leading to national economic development (Figure 10.4).

Arguments in Favor of the Model

- Foreign investors can start the operation with low fixed cost initially as natural resources are ready inputs here. Since Bangladesh is endowed with natural sites, there is no requirement to extract and import natural resources.
- Foreign units have the opportunity to work with local people. The benefits lie in the experience and prior knowledge of the people regarding the place. Nonetheless, the variable cost will be lower for cheap labor.
- The number of domestic tourist is increasing. In addition, per capita income exhibits an upward trend. Domestic tourists will get the opportunity to reap the benefits of tourism at own area and at lower transaction cost. This implies quick recovery of fixed and variable costs for foreign investors.
- The geographical location makes for easy access to Bangladesh. So, the number of foreign tourists is expected to increase. The foreign unit will get a large market including domestic and foreign consumers.
- The smooth backward linkage accelerates the operations of the foreign units (Figure 10.5).
- The existing foreign investors have faced heterogeneous blocks while working in manufacturing sector. The tourism sector is different in characteristics than that of manufacturing. Hence, it will create a positive image to foreigners.
- The model will operate viewing the new concept PLPFP (Public-Local Private-Foreign Partnership). More success in predicted as the term

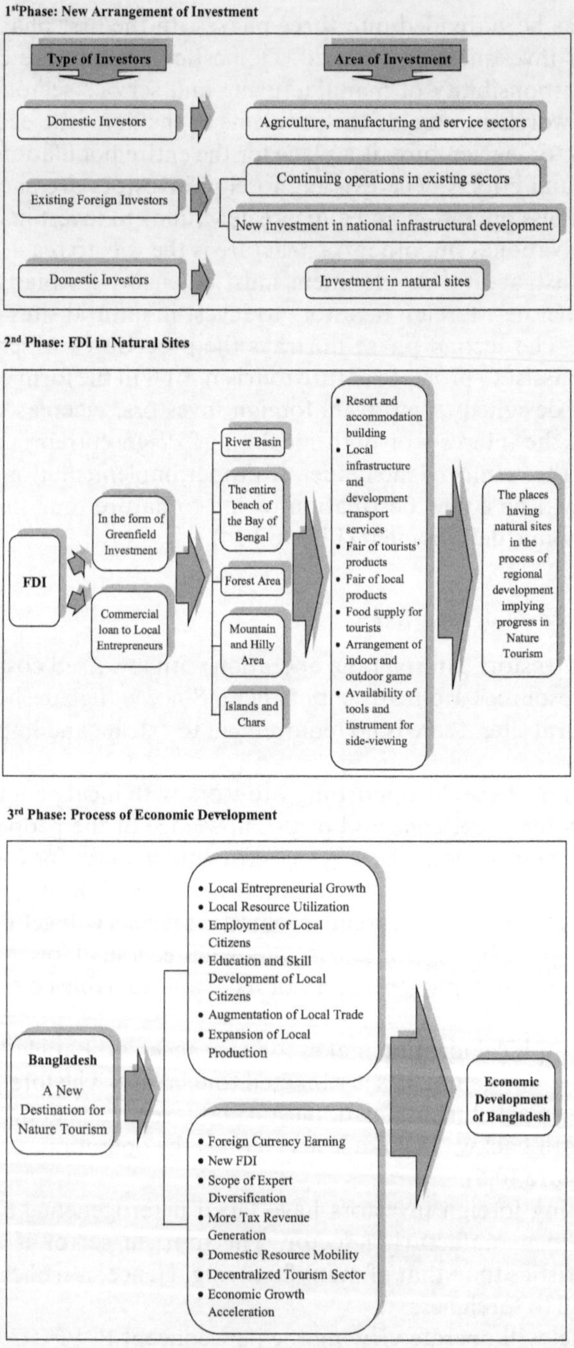

Figure 10.4 Prospects of FDI in nature-tourism (PFT model).

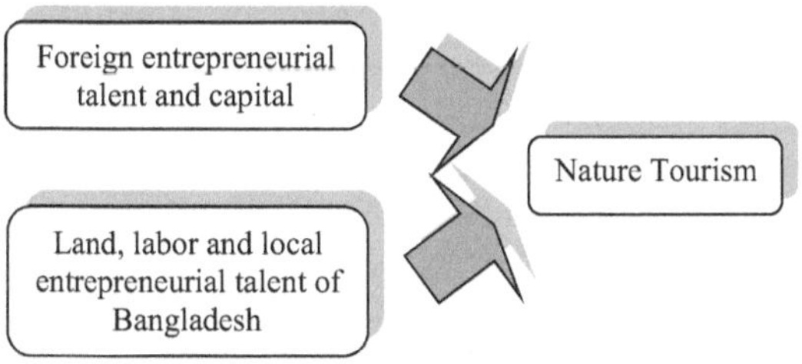

Figure 10.5 Smooth backward linkage in nature tourism.

PLPFP suggests. In the public sector, the Bangladeshi government will liberalise the protocol and procedure both for foreign tourists and investors. The local private sector will provide consultancy for site development. Finally, the foreign firm will lead the entire function with capital accumulation.

- At present, the tourism sector is monopolised. The PFT model is expected to introduce monopolistic competition which is helpful to rescue the economy from dead-weight loss.
- Bangladesh shows enthusiasm in attracting foreign investment in the sectors which need to be developed. In order to achieve long run economic development, the tourism sector attains the eligibility for investment and contribution to economy.

Challenges

The proposed PFT model is going to face the following challenges:

- Human intervention in nature may cause harm to biodiversity
- Man-made structures in natural sites may cause damage to environment. Damage can also occur from tourists' activities
- Construction activities imply threats for workers' health
- Vicious circle of pirates in the Bay of Bengal may undermine the pleasure of tourists

Recommendations

The following recommendations have been suggested to overcome the challenges.

- Existing laws and rules should be implemented and monitored properly to avoid the loss of biodiversity
- Environment-friendly equipment and architectural design may mitigate environmental damage. Tourists' awareness will help protect the

environment. Local volunteers can support the waste management program taken by authorities
- Implementation of security and safety measures for workers should be prioritised
- The government should negotiate with local bodies, neighboring countries, and international agencies to eradicate piracy

Conclusion

In order to stimulate nature tourism growth and economic development, FDI is a significant factor. The investment is being received by Bangladesh with gratitude. Therefore, policies and policy changes are constantly implemented in order to maintain the incentive. The study has started the journey with the research question of the potential of nature tourism and the prospects of FDI in nature tourism in order to attain economic development. To answer the question, the pattern of FDI in Bangladesh and potential of nature tourism have been explored. The FDI trend portrays the feasibility of Bangladesh to receive FDI. The country can attract more FDI owing to low labor cost, strategic geographical location, liberalisation policy, and non-discriminatory attitude. Moreover, the exploration of natural sites illustrates the necessity of investment with satisfactory returns. Hence, the tourism sector, particularly nature tourism can be a lucrative destination for FDI. The study has developed a model to show how FDI can help attain nature tourism growth and economic development. The model, if implemented as indicated, can make the country an attractive destination both for tourists and FDI, leading to accelerated economic development. In the end, the study suggests that liberalised investment policy and government intervention can run the operations smoothly. Moreover, international corporations can be a good means to help implement this model.

References

Ali, M 2004, *Bangladesher Sandhanay*, 3rd edn, Student Ways, Dhaka.

Alauddin, M, Shah, MGH & Ullah, H 2014, 'Tourism in Bangladesh: a prospects analysis', *Information and Knowledge Management*, vol. 4, no. 5, pp. 67–73.

Bangladesh Bank 2019, *Foreign Direct Investment in Bangladesh*, Survey Report.

Baker, JC 1998, *International Finance: Management, Markets and Institutions*, Prentice-Hall, New Jersey.

Blanke, J & Chiesa, T 2007, The travel and tourism competitiveness report 2007: furthering the process of economic development, *World Economic Forum*, viewed 5 May 2021, https://citeseerx.ist.psu.edu/viewdoc/download?doi=10.1.1.178.3367&rep=rep1&type=pdf.

Dey, S & Awal, BH 2017, 'Impacts of foreign direct investment on economic growth of Bangladesh: an economic exercise', *Asian Business Review*, vol.7, no. 2, pp. 71–78.

Hafsa, S 2020,'Economic contribution of tourism industry in Bangladesh: at a glance', *Global Journal of Management and Business Research (F)*,vol. XX, no. 1, version 1, pp. 29–38.

Hasan, MM & Hossain, MR 2014, 'Development of tourism industry through SME: a study on Comilla', *International Journal of SME Development*, vol. 1, no. 1, pp. 59–76.

Hossain, MR 2021, 'Inward foreign direct investment in Bangladesh: do we need to rethink about some of the macro-level quantitative determinants?', *SN Business of Economics*, *1*, Article number: 48, viewed 10 May 2021, https://www.macrotrends.net/countries/BGD/bangladesh/foreign-direct-investment.

Islam, F & Islam, N 2004,'Tourism in Bangladesh: an analysis of foreign tourist arrivals', *Journal of Travel and Tourism Research*, vol. 4, no. 1, pp. 26–39.

Jahan, N & Amin, MR 2014, 'Sustainable tourism development in Bangladesh: an empirical study on Sylhet', *Journal of Business Studies*, vol. 23, no. 2, pp. 239–260.

Manzoor, SH & Chowdhury, ME 2017,'Foreign direct investment in Bangladesh: some recent trends and implications', *Journal of Business & Economic Research*, vol. 15, no. 1, pp. 21–31, viewed 16 May 2021, https://clutejournals.com/index.php/JBER/article/view/9855.

Mondal, WI 2003, 'Foreign direct investment in Bangladesh: an analysis of perceptions of prospective investors', *Studies in Economics and Finance*, vol. 21, no. 1, pp. 105–115.

Shamsuddoha, M 2004,'Sustainability of tourism industry: Bangladesh perspective', for sustainable exploitation of minerals and natural resources held in IIT conference, Kharagpur, India, viewed 15 May 2021, https://papers.ssrn.com/sol3/papers.cfm?abstract_id=1314779.

Sultana, S 2016, 'Economic contribution of tourism industry in Bangladesh', *Journal of Tourism, Hospitality and Sports*, vol. 22, pp. 45–54.

UNCTAD 2020, *World Investment Report*. United Nations Publications, New York, USA.

World Bank 2021, *Foreign Direct Investment, Net Inflows (BoP, Current US$) – Bangladesh*, The World Bank, viewed 12 May 2021, World Bank https://data.worldbank.org/indicator/BX.KLT.DINV.WD.GD.ZS?locations=BD.

WTTC 2017, *The Economic Impact of Travel and Tourism*, World Travel and Tourism Council, viewed 7 May 2021, https://wttc.org/Research/Economic-Impact. World Travel and Tourism Council.

WTTC 2019, *The Economic Impact of Travel and Tourism*, World Travel and Tourism Council, viewed 11 May 2021, https://wttc.org/Research/Economic-Impact. World Travel and Tourism Council.

11 Effect of Foreign Direct Investment in Tourism on Economic Growth

Empirical Study in Indonesia

Bryna Meivitawanli

Introduction

Indonesia is a country with a vast amount of culture, ethnic diversity, languages, islands and natural resources. It is the fourth largest country in the world in terms of population (World Bank 2021). Many often mistake Indonesia to be a Muslim country. Although, it has the largest Muslim population in the world, Indonesia is not governed by Islamic law, thus it cannot be considered as a Muslim country. The country's official motto is *Bhinneka Tunggal Ika* or Unity in Diversity. There are more than 700 living languages (Central Intelligence Agency 2021) and 1300 ethnic groups in Indonesia (Yuniarni 2016). Since the society is very diverse, local people are friendly to foreigners and other people in general. Diverse ethnic groups also mean a wide variety of local cuisines. Traditional foods from different parts of Indonesia are rich in spices and herbs which was one of the main reasons the Dutch colony started back in the early 17th century. Not only is its society diverse, Indonesia also ranks third in the world for its biodiversity (Butler 2016). Indonesia lies across the equator; therefore, it is a tropical country with sun shining all year round. It experiences only two seasons, wet and dry.

Due to all of the characteristics mentioned above, Indonesia is a suitable tourist destination for many kinds of people at any time of the year. The most famous island to go to is Bali. The popularity of Bali exceeds Indonesia itself, many foreigners recognise Bali but not Indonesia. This is both fortunate and unfortunate since Indonesia has many more to offer apart from Bali. Jakarta placed fifth as the world's fastest growing cities in terms of 2017 to 2018 direct travel and tourism GDP (World Travel and Tourism Council 2019). Due to its popularity abroad and large population of local tourists, Indonesian economy is highly affected by its tourism industry especially with regards to foreign currency earnings. Inbound tourists mainly come from China, Singapore, Malaysia and Australia in 2018 (Organisation for Economic Co-operation and Development 2020). In 2019, international tourist arrivals were recorded at 16.1 million, an almost 2% growth from 2018 (Akhlas 2020). Tourism direct GDP has been stable at around 4% out of total GDP. Inbound tourism has increased overtime until 2019, so has the

DOI: 10.4324/9781003155492-14

total number of establishments and number of employees employed in the tourism industry (World Tourism Organization 2020).

According to Travel and Tourism Competitiveness Report 2019 published by World Economic Forum, Indonesia ranks 40th out of 140 countries in terms of travel and tourism competitiveness index (TTCI). In Asia Pacific, Indonesia ranks 12th out of 22 countries and is considered as the most improved in health and hygiene. Out of four subindexes of TTCI, enabling environment, T&T policy & enabling conditions, infrastructure and natural & cultural resources, Indonesia performs well in the second and last subindexes. In its effort to nurture the tourism industry, the Indonesian government developed ten priority programs, which are digital tourism, homestay, air accessibility, branding, top ten origination, top three main destination, ten new priority tourism destinations, certified human resources in tourism and tourism awareness movement, tourism investment growth and crisis centre management. This initiative is supported by FDI in the tourism industry of around USD 13.3 billion which is expected to be a new driver of the country's economic growth (Kumairoh 2019).

This chapter discusses FDI in Indonesia's tourism industry and its effect on the nation's economic growth. As can be seen, Indonesia's tourism sector is growing fast and it has become one of the most important sectors for economic growth. Despite common belief and hearsay, it is unclear whether FDI in this industry actually contributes to economic development. Therefore, empirical study in Indonesia is conducted in this regard. Travel and tourism is one of the hardest hit sectors with the ongoing COVID-19 pandemic. Countries all over the world have closed their borders and restricted travels between countries as well as within country. Indonesia is no different. Therefore, data from 2020 is not included in this research due to it being an anomaly. The research aims to give an insight onto the significance of FDI in tourism industry (TFDI) in fostering Indonesia's economic growth and whether there are necessary supporting factors or absorptive capacities to actualise the effect of TFDI on economic growth. The chapter is divided into sections discussing Tourism in Indonesia, TFDI in Indonesia, Literature Review, Methodology, Results and Discussion and Conclusion, Recommendation and Limitation.

Tourism in Indonesia

As mentioned previously, there is much that Indonesia can offer to tourists. Starting from nature, including beaches, ocean, mountains, volcanoes, lakes, craters, plateau, waterfall and so on, Indonesia also has many cultural sites, such as Buddha temples, Hindu temples, Confucianism temples, mosques, churches, archaeological sites, traditional houses and many more. Those who prefer luxury can enjoy high quality spas, golfing, shopping, amusement parks as well as nightlife. Tourists can indulge in a wide variety of cuisines during their stay apart from enjoying other tourist attractions and activities. Each of these four types of leisure is described in this section, but it is important to note that these are only some of what Indonesia has to offer.

Nature Tourism

Indonesia is the largest archipelagic country in the world. Its coastline spreads as long as 54,716 kilometres. More than half of the land area is covered by forest. It is the second most heavily forested area in the world after the Amazon. It is also country with the most volcanoes in the world amounting to 76 historically active volcanoes (Central Intelligence Agency 2021). Due to its geography, Indonesia is the right destination for nature tourism. Bali is certainly known for its natural beauty. White sandy beaches, beautiful hills and magnificent rice terraces can be found easily. But Bali is not the only one, there are many other areas with landscapes as stupendous as Bali. Several examples are Lombok, Bintan, Karimunjawa, Wakatobi and so on.

Divers and surfers can consider Indonesia as one of their ultimate destinations. The highest record of marine life diversity in the world is located in Raja Ampat (Conservation International n.d.) boasting over 1,300 species of coral reef fish, 700 species of mollusk, 75% of the world's species of hard coral, 13 marine mammal species and many more (World Wide Fund for Nature n.d.). Scuba diving in Indonesia is not only breathtaking but also affordable. Marine life such as manta rays, dolphins, turtles, sharks and many others can be found in Indonesian water. Moreover, Indonesia also has ship wrecks to discover. Tourists who are more interested in terrestrial animals instead of aquatic can also discover numerous exotic and rare animals in Indonesia. Examples are Komodo dragon, Anoa, Javan rhinoceros, Sumatran harimau, Babirusa, Cendrawasih and so on (Hananto 2020). Indonesia is also home to the largest flower in the world, *Rafflesia arnoldii* (Andrew 2020).

As mentioned previously, Indonesia hosts a large number of volcanoes, both active and inactive. Apart from the hike and view from the top, many of the volcanoes have unique craters. Among them are the White Crater or Kawah Putih in West Java and crater lake Kelimutu which comprises of three lakes with three different colours. The lake colours change from time to time making it even more interesting for visitors. Another lake in North Sumatra called Lake Toba was created by a supervolcano eruption and has become a famous tourist destination for local and foreign tourists.

Cultural Tourism

Majority of Indonesian are Muslims, therefore many mosques can be found in cities all over the country. One of the most notable is Istiqlal Mosque in Jakarta, which is the largest mosque in Southeast Asia (Enjoy Jakarta n.d.). Interestingly, it is located right across St. Mary of the Assumption

Cathedral. On Fridays, Muslims who come to perform Friday prayers use the Cathedral's parking space while on Sundays, Christians use the mosque's parking space. This is one small example of religious tolerance in Indonesia. Moreover, Indonesia hosts numerous Hindu and Buddhists temples. Borobudur temple is the largest Buddhist temple in the world (Guinness World Records 2021). It is located in Central Java, not far from Prambanan Temple which is the largest Hindu temple in Indonesia. Both temples are recorded in UNESCO World Heritage list (United Nations Educational, Scientific and Cultural Organization n.d.). As can be seen, Indonesia is the right destination for cultural tourism especially with regard to various religions.

Indonesia is very varied in terms of ethnic groups. Visitors can visit Beautiful Indonesia Miniature Park to get a glimpse of Indonesian cultural heritage. The park includes traditional houses, places of worship, costumes, flora, fauna, museums and a manmade lake as a miniature of Indonesian major islands (Yosephine 2018). As the name suggests, the park is a miniature of the whole country so that tourists can experience different cultures of the 34 provinces in Indonesia without having to spend time and money to visit each one of them. Due to the large number of ethnic groups, there are different customs and languages from province to province. Out of all the provinces, Bali is the only province with Hindu as the major religion. Therefore, Bali is not only famous for its natural beauty but also strong Hindu customs. Many local and international tourists are attracted to come and visit the Hindu temples and to experience Balinese culture. Hinduism and local Balinese culture are very strong and can be seen in every corner of Bali including its traditional dances, music, arts, architecture and rituals.

Javanese culture also has its own sense of authenticity and appeal. Javanese traditional dances, music and various forms of arts are well loved all over the world. One of Javanese most highly developed forms of art is batik. It is an ancient tradition of hand-printing textiles by coating with wax the parts not to be dyed (Merriem-Webster n.d.). Tourists both local and foreign love to buy Batik in many forms including cloths, clothes, bags, wallets, fans and many others as souvenirs. Indonesian Batik is listed in UNESCO list of Intangible Cultural Heritage of Humanity. There are many places especially in Yogyakarta that offer lessons on how to batik. Apart from religions and ethnicities, foreigners can learn more about Indonesian history and the legacy of the colonial era through museums spread all over the country.

Luxury Tourism

Indonesia is a developing country, however its capital city and some of the large urban areas are very well developed with skyscrapers scattered all around. Five-star international hotel chains can be found easily such as Four Seasons, Ritz-Carlton, JW Marriott, Hilton, Fairmont, Shangri-La and so on. There are also many world-class local hotels and resorts, especially in Bali. Indonesia was also awarded the best spa destination in the world in

International Wellness Awards 2012 (CNN Indonesia 2017). Balinese and Javanese traditional spas have their own uniqueness and competitive advantage. Traditional herbs are used both as part of the treatment as well as additional service in the form of drinks or food. Jamu which are traditional drinks made of different kinds of herbs are commonly consumed by the locals for health purposes. Many have been proven to boost immunity and provide other health benefits.

Other than spa, Indonesian Ministry of Tourism and Creative Economy is pushing golf as one of the focus of Wonderful Indonesia, the country branding for its tourism industry. In 2018, Wonderful Indonesia Golf Tournament was held to promote Indonesian mountain and highland tourism. Golf courses with international quality and beautiful landscapes can be found in many places around the country. Rainforest jungle, active volcano, ocean, river, limestone cliffs, rice paddies and lake are some of the natural surroundings that can be enjoyed while golfing in Indonesia.

Shopping and nightlife are also part of luxury tourism that people enjoy. These two are well accommodated in Indonesia. There are many luxurious shopping malls and complexes with designer brands filling up the outlets. High-class bars and clubs are mostly opened in the large cities of Indonesia. Since Indonesia is a Muslim majority country, alcoholic beverages are not sold in mini markets or supermarkets, but they can be bought easily in bars, clubs, hotels and resorts. Bali is also different from the rest of the country since the majority of the population is Hindu; therefore many limitations such as alcoholic beverages, pork and lard cuisines as well as proper attire are not restricted in Bali. Most Indonesian people can speak at least basic English since Indonesian language also use Latin script from A to Z and English are taught in schools. This is especially true in Bali since local people encounter foreigners on a daily basis.

Culinary Tourism

Indonesia is not only famous for its natural beauty and cultural heritage but also its food. The first and second most delicious food in the world picked by CNN readers both come from Indonesia (Cheung 2017). The first being Rendang and Nasi Goreng follows behind. Another food also made it into the top 50 list which is Satay at 14th place. This does not come as a surprise since as mentioned before, Indonesia was first colonised by the Dutch due to its spices and herbs. Certainly, Indonesian food explodes with flavours coming from those spices. Rendang requires a long time to cook, it is beef simmered in coconut milk with additions of lemongrass, garlic, galangal, turmeric, ginger and chillies (Cheung 2017). It is cooked for hours in order to create a tender, aromatic and flavourful dish. Nasi Goreng directly translates to fried rice. Although it sounds ordinary, Indonesian Nasi Goreng is well-known all over the world. It is normally served with egg, chicken and prawn crackers; a delight that should not be missed by anyone visiting the country.

Each ethnic group or province has their own traditional food. These foods are very well loved by the local people. The one thing that Indonesians who live abroad miss the most about their home country is its food. Tourists with different kinds of taste can find local dishes suitable to their taste due to the large variety of dishes Indonesia has to offer. Another example of how well accepted Indonesian food internationally is through the success of Indomie, instant noodle produced by Indonesian company, Indofood. Indofood has become the largest instant noodle producer in the world with 16 factories in several countries. Its factory in Nigeria is the largest in Africa. Furthermore, Indomie has been exported to 60 countries (Indonesian Ministry of Trade 2018). Apart from main dishes, Indonesia also has numerous fruits that are of high quality but affordable. Several examples include durian, Bali pomelo tangerine, rambutan, salak and many others (IPB 2016).

TFDI in Indonesia

Inward FDI to Indonesia is highly encouraged by the government including FDI into the tourism industry. Two sectors in Indonesia are especially attractive for both domestic and foreign investors. These two are tourism and lifestyle sector (Indonesia Investment Coordinating Board 2018). Indonesia's tourism industry is growing from year to year and the government is trying to develop this sector even more through foreign investments. Therefore, TFDI receives close attention from the government. In this section, issues, challenges and prospects of FDI in Indonesia's tourism industry are discussed.

Issues

As mentioned previously, TFDI is encouraged by the Indonesian government. However, the number of projects as well as value of investments in the services sector and hotels and restaurants in particular has been fluctuating. There is no clear trend over the years. Figure 11.1 below shows the fluctuation from 2002 to 2019 using data from Statistics Indonesia which is Indonesia's central statistics bureau. As it can be seen, it is very different compared to stable increase in trend of foreign visitor arrivals to Indonesia from the same time period. Looking at the two line-graphs, it shows that FDI in hotel and restaurant sector does not seem to have an immediate effect on tourist arrivals.

This raises an issue of whether TFDI actually helps the Indonesian tourism industry since it is not reflected in the number of foreign tourist arrivals. That is also the main reason why this research is conducted in the first place. However, this research studies the effect of TFDI on economic growth in

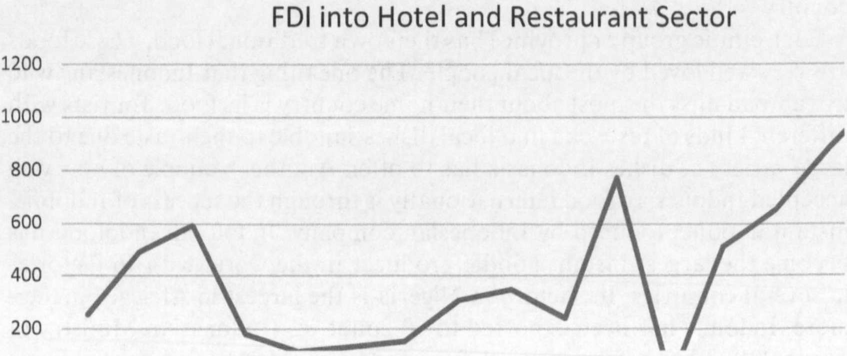

Figure 11.1 Foreign direct investment realisation (value) in the hotels and restaurants sector from 2002 to 2019.

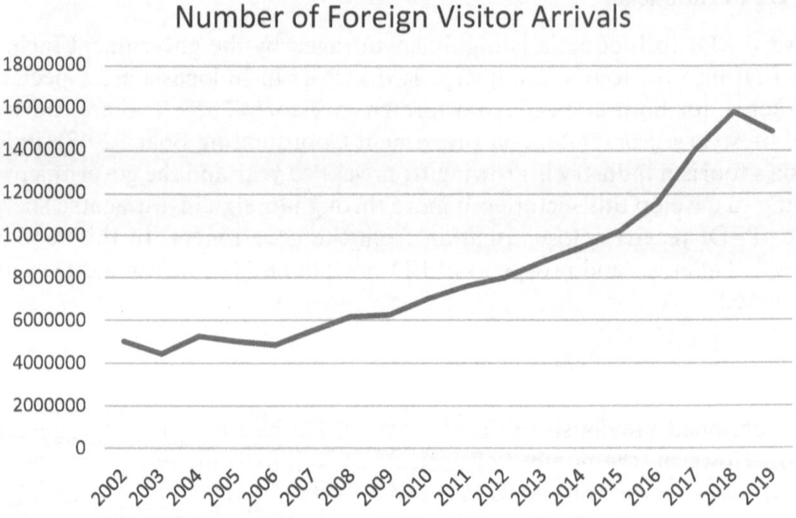

Figure 11.2 Number of foreign visitor arrivals to Indonesia from 2002 to 2019.

general. Therefore, another figure is shown below to show the trend of total GDP by expenditure as well as GDP from the restaurants and hotels sector.

Chart above shows that both total GDP and GDP from restaurants and hotels are increasing steadily. Once more, it shows that data pattern of TFDI and GDP are not similar. This raises question of the benefits that FDI brought to the host country's economy which is Indonesia. A more in-depth data analysis is performed in this research to examine whether TFDI affects economic growth. This is a very important issue since the Indonesian

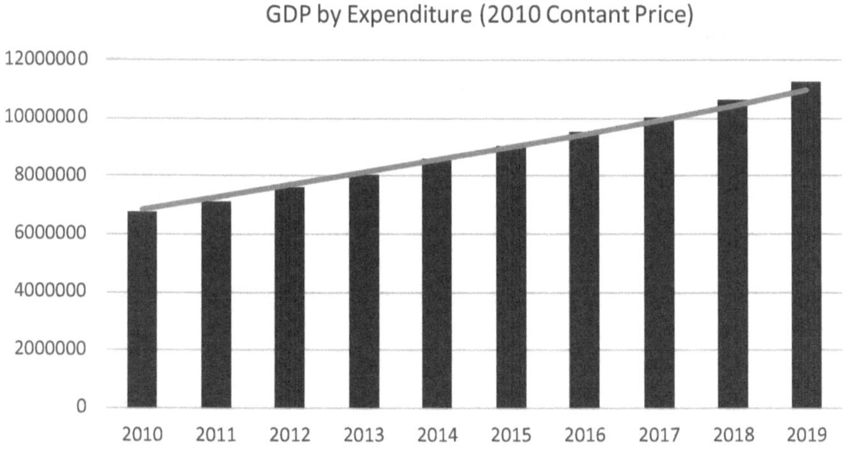

Figure 11.3 Total real GDP and real GDP from restaurants and hotels from 2010 to 2019.

government encourages TFDI. It means resources used to attract TFDI can be better allocated to other sectors or aspects that actually bring positive effect on the economy. Indonesia is a developing country; therefore it should wisely distribute the limited budget the government has.

Challenges

In the current period, the year of 2021, the COVID-19 pandemic that is still active all over the world with the second and third waves hitting many countries is the biggest challenge for the tourism industry. Although vaccination has been pushed by numerous governments worldwide, however the virus is continuously mutating into different varieties causing more and more damage. The tourism industry is certainly one of the hardest hit sectors due to travel restrictions and health concerns. Indonesia has been hit as hard as the other neighbouring countries. Indonesia recorded the highest number of cases among ASEAN countries due to its large population, however in terms of percentage of cases from total population, Malaysia, Singapore and the Philippines are higher than Indonesia. This is a challenge that the tourism industry worldwide is facing, not just Indonesia. Governments are doing their best to combat this virus, and hopefully the tourism industry can recover steadily when herd immunity is reached. In this research, this challenge is excluded from the analysis due to its magnitude by using data up until 2019, before the pandemic spread in Indonesia.

Apart from the COVID-19 pandemic, there are challenges specific to Indonesia such as lack of infrastructure and environment sustainability.

Indonesia's score for these two subindexes along with health and hygiene is considered low when compared to other countries in Asia-Pacific. In the case of health and hygiene, Indonesia is appraised as the most improved country in the region; therefore, this should not be a significant challenge in the future. However, infrastructure as a whole as well as tourist service infrastructure and environment sustainability requires a lot more dedication from the government as well as private sector. These two are common challenges in this region. TFDI can help with the development of infrastructure. However, environment sustainability is one of the concerns regarding TFDI. The Indonesian government has not included provisions related to environment sustainability in its agreements (Widiatedja 2019). Therefore, foreign investors are not entitled to sustain Indonesian natural resources. This should be amended by the government in order to make sure that its tourism industry is sustainable for years to come.

Prospects

Since the Indonesian government tries to attract as much TFDI as possible in order to accomplish the goal of creating ten new Bali(s), many regulations as well as the investment environment have been improved. As evaluated by UNTWO in its TTCI, Indonesia is doing especially well in travel & tourism policy and enabling conditions. Indonesia scored 5.9 for prioritisation of travel and tourism, the second highest in South-East Asia after Singapore (World Economic Forum 2019). It means that the efforts expended by the government are not worthless and this is reflected through the index. This is good prospect for foreign investors since there are leniencies and incentives given to those who are investing in the travel and tourism industry. This also means that tourist arrivals should be increasing due to the attention the government has given to this sector. Investors can reap higher profit as the market grows larger.

The government initiatives to create ten new Bali(s) is definitely a bright prospect for Indonesia's tourism industry as well as TFDI. It opens up many possibilities and opportunities for investment in different aspects such as hotels, restaurants, infrastructure, amusement parks, golf course and so on. There are also ten different destinations to choose from which can suit various kinds of investors, such as those specialised or interested in nature tourism, luxury tourism, culinary tourism, sport tourism or others. The destinations selected by the government are already well-known with the local tourists, therefore investors are not taking huge risks since Indonesian local tourists is a large market due to the sizeable population. These ten destinations are Lake Toba, Tanjung Kelayang Beach, Tanjung Lesung, The Thousand Islands, Borobudur Temple, Mount Bromo, Lombok Mandalika, Labuanbajo, Wakatobi and Morotai Island (Invest Islands 2019).

As shown previously, the number of foreign visitor arrivals to Indonesia has been steadily increasing over the years although there is a decrease in 2019. This is expected due to the COVID-19 pandemic. This along with the government initiatives show that there is still large room for the industry to grow. As highlighted in the TTCI 2019, Asia-Pacific is rapidly improving travel and tourism region with abundant natural and cultural resources (World Economic Forum 2019). Air transportation is also booming with more and more routes served by an increasing number of airlines. Therefore, apart from the growing tourism industry as a whole, Indonesia also has the chance to attract international tourists who are visiting other ASEAN countries due to proximity and uniqueness.

Literature Review

FDI is not a new phenomenon and has been extensively studied in the past. One of the most commonly discussed topics with regards to FDI is its contribution to economic growth. Many developing countries try to encourage FDI in order to boost their economies. However, it is not always the case when actual data are analysed. Numerous empirical researches do not come to the expected conclusion due to many factors. An example is the need of absorptive capacities as prerequisites for host countries in order to effectively and efficiently absorb the benefits of FDI and realise it in the form of economic growth. These literatures are discussed in this section.

Foreign Direct Investment

FDI is defined as the objective of establishing a lasting interest by a resident enterprise in one economy, direct investor, in an enterprise, direct investment enterprise, that is resident in an economy other than that of the direct investor (Organisation for Economic Co-operation and Development 2018). Empirical research suggests conflicting results depending on the dataset. The varying results can be due to different country characteristics, different industries taken into account, different time frame of the dataset and many others. Therefore, conducting empirical research on a certain industry in a specific country with the latest data is beneficial.

Lindblad (2015) concluded that the nexus between FDI and economic growth in Indonesia is less straightforward compared to neighbouring countries. Increase in economic growth in the past was not induced by FDI but by oil boom and industrialisation. FDI was seen as an economic growth driver in the 1990s; however, it did not last for a long time due to the Asian financial crises. This shows that country specific study is important and generalisation of a certain group of similar countries is unwise. Furthermore, a research by Khaliq and Noy (2007) studied the impact of FDI inflows to different sectors in Indonesia. The research discovered that although the aggregate effect of FDI on economic growth is positive, but different sectors were impacted differently by FDI. The impact varies from significant positive results, insignificant results and even significant negative results. This shows that sectoral level study is important in a single country analysis.

Thus, performing a single sector study on a single country increases not only the theoretical contribution but also practical contribution of the research. This research studies the effect of FDI in tourism industry on Indonesian economic growth in the last two decades, excluding the year 2020 due to COVID-19 pandemic.

Tourism Foreign Direct Investment on Economic Growth

TFDI has not been studied as extensively as aggregate FDI, but it has received a lot of attention in recent years. Similar to aggregate FDI, research on TFDI arrived at different conclusions. It is important to find out the actual effect of TFDI on economic growth since the government of Indonesia is actively encouraging foreign investment into the tourism sector without sound confirmation that the local economy instead of only foreigners benefitted from it. There are damages that TFDI brings to the host country such as environmental damage. Foreign hotels tend to be bigger; therefore, it uses more energy and more water compared to local hotels (Barrowclough 2007). These possible harms should not be welcomed if ultimately there is no significant positive effect of TFDI on the local economy.

A panel study of 18 OECD countries found that TFDI has no significant effect on economic growth (Meivitawanli 2018). On the other hand, a single country analysis on Mauritius found that TFDI significantly contributes to economic growth (Fauzel et al. 2017). A panel study of ASEAN countries also found that TFDI is one of the factors inducing economic growth (Pratoomchat 2017). More interestingly, a recent study in Morocco found that TFDI has negative effect on economic growth using data from 1983 to 2018. In the same research, it was found that FDI in non-tourism sector has significant positive effect on economic growth (Menyari 2020). This stresses the importance of conducting a research that is specific to TFDI since it does not necessarily behave in the same manner as FDI in other sectors.

Many other literatures focused on the impact of TFDI on tourism growth instead of economic growth. Some found that TFDI does positively affect tourism development (Fauzel 2021; Rashid et al. 2020; Al hallaq et al. 2019), while many are not empirical studies. Therefore, it can be seen that there is limited study on the effect of TFDI on economic growth and the current body of knowledge is not in agreement of its effect. Impact of aggregate FDI on economic growth has been a long-standing debate as well. In order to better understand the variances in results, many researchers take several absorptive capacities into account. These factors can play a role in the nexus between TFDI and economic growth and therefore are discussed in the next subsection.

Absorptive Capacities

There are many factors that play a role in the relationship between FDI and economic growth. Numerous literatures have studied varying factors that become prerequisites for the host country in order to take in positive externalities FDI brings into the country. These factors include human capital (Borensztein et al. 1998; Li & Liu 2005; Calvo & Sanchez-Robles 2003), financial market development (Alfaro et al. 2004, 2010; Azman-Saini et al. 2010), trade openness (Iamsiraroj 2016; Iamsiraroj & Ulubaşoğlu 2015; Balasubramanyam et al. 1996), economic stability (Calvo & Sanchez-Robles 2003; Gbakou et al. 2009) and many others. Certainly, these factors were not always found to be significantly related to the nexus. However, much empirical research has found that they play an important role in making sure that the host country economy acquires the benefit of aggregate FDI.

Unlike aggregate FDI, there is not much research that takes absorptive capacities into account when examining the effect of TFDI on economic growth with the exception of a cross-country study done by Meivitawanli (2018). It was found that trade openness and financial market development are not the appropriate absorptive capacities for the 18 OECD countries studied. This is a gap in the literature that can be filled by this research, especially since this is a single country analysis instead of a cross-country study. Furthermore, Indonesia is a developing country that has a different level of human capital and trade openness compared to the developed countries previously studied. Based on the literature review, it can be seen that this study fills in research gaps in the current body of knowledge.

Research Methodology

This study is an empirical research using quantitative methodology. Quantitative research is quantifying and analysing variables to get results that involves utilisation and analysis of numerical data using specific statistical techniques (Apuke 2017). This research uses secondary data taken from reliable sources such as the Indonesian central statistics bureau. The indicator of each variable as well as the data source of each indicator is explained in this section. Before that, explanation of the specific research methodology used in this research is provided. The method used to study the relationship between TFDI and economic growth is least squares regression.

Least Squares Regression

Regression analysis is commonly used in the literature. It uses the conceptually logical process of using an equation to express the relationship between dependent and independent variables (Montgomery et al. 2021). Simple linear regression involves one dependent and one independent variable, whereas multiple regression is an extension of simple linear regression. In multiple regression, there is one dependent variable and several independent variables. In this case, the interaction terms between TFDI and absorptive capacities are introduced later on into the equation. Therefore, the appropriate methodology is multiple regression since there is more than one independent variable.

The data are analysed using the least squares regression method. In order to run multiple regression, the hypotheses are shown as regression equations. There are four equations to be tested. These four are listed below.

$$EG_t = \alpha + \beta_1 TFDI_t + \beta_2 HC_t + \beta_3 TR_t + \beta_4 FD_t + \varepsilon_t \tag{11.1}$$

$$EG_t = \alpha + \beta_1 TFDI_t + \beta_2 HC_t + \beta_3 TR_t + \beta_4 FD_t + \beta_5 TFDIHC_t + \varepsilon_t \tag{11.2}$$

$$EG_t = \alpha + \beta_1 TFDI_t + \beta_2 HC_t + \beta_3 TR_t + \beta_4 FD_t + \beta_5 TFDITR_t + \varepsilon_t \tag{11.3}$$

$$EG_t = \alpha + \beta_1 TFDI_t + \beta_2 HC_t + \beta_3 TR_t + \beta_4 FD_t + \beta_5 TFDIFD_t + \varepsilon_t \tag{11.4}$$

EG represents economic growth, α is the intercept, β is the coefficient of respective independent variables, TFDI stands for FDI in the tourism industry, HC stands for human capital, TR represents trade openness, FD stands for financial development, ε is the error term and subscript t represents the time dimension since the analysis is done using time series data. The second, third and fourth equations introduce new variables which are the interaction terms of TFDI with three absorptive capacities, i.e., human capital, trade openness and financial development. The interaction term determines whether these factors are needed to render the positive effect of TFDI as significant. Statistical software used for data analysis is Eviews 11.

Indicators and Data Sources

In total, there are five variables that are tested in this research: one dependent variable and four independent variables. The dependent variable is economic growth. There are many indicators used by previous literature. Most of them are a variation of GDP growth. In this research, economic growth is proxied by annual GDP growth taken from the World Bank. In the case of TFDI or FDI in tourism industry, the indicator for this variable is FDI in hotels and restaurants (Samimi et al. 2013; Steiner 2010). The indicator for HC is the school participation rate from 16 to 18 years old. This is the age where teenagers are supposed to attend senior high school. Since Indonesia is still a developing country, the level of education completed by the population in general is not very high. Therefore, taking senior high school should show variation and improvement of human capital over the years. The government used to have a program of compulsory nine years of education; however, in 2016, it has been changed to 12 years to include senior high school (Ghofar 2015). In the case of trade openness, the indicator used is total trade out of GDP as has been used by many literatures in the past (Batten & Vo 2009; Carkovic & Levine 2002; Zekarias 2016). Financial development is proxied by domestic credit to private sector by banks as percentage of GDP (Khan 2007; Kholdy & Sohrabian 2005). The sources of these indicators are taken from Indonesian official central statistics bureau called Statistics Indonesia or Badan Pusat Statistik (BPS) and World Bank. TFDI indicator requires further calculations; the raw data are taken from Statistics Indonesia and the World Bank. The data are annual data from 2000 to 2019 comprising 20 years in total.

Results and Discussion

The results of least squares regression are shown in Table 11.1. There are four regressions to test the four hypotheses. The standard errors shown are robust to heteroskedasticity. The Durbin-Watson statistics which test for autocorrelation are 1.66, 1.80, 2.19 and 1.95 for the first, second, third and fourth equations, respectively. These results are all between the acceptable range of 1.5 and 2.5 (Garson 2012). The R square result of the first equation is 0.56, which means that over half of the variation of economic growth is explained by the variation of the four independent variables. The R square results of the second, third and fourth equation are 0.61, 0.67 and 0.64, respectively. The addition of interaction term increased the R square by at least 5% with TFDITR giving the highest increase of more than 10%.

As can be seen, Table 11.1 shows that all equations have significant F statistics since the probabilities are all well below 0.05. It means that all four equations are significant as a whole. All equations are run using data from

Table 11.1 Least squares regression results

Explanatory variables	*Dependent variable: EG*			
	(1)	**(2)**	**(3)**	**(4)**
Number of obs.	20	20	20	20
Prob (Wald F-stat)	0.00000	0.00000	0.00000	0.00000
TFDI	−30.59520**	2093.933	−2004.680*	965.9145*
	(5.671214)	(1377.439)	(791.6464)	(350.5875)
HC	−0.118793*	−0.069371	−0.101809*	−0.100099*
	(0.043099)	(0.047519)	(0.041445)	(0.041794)
TR	0.005596	0.008048	−0.016607	0.008328
	(0.005596)	(0.024626)	(0.032023)	(0.024488)
FD	0.202570**	0.178264*	0.166994*	0.214336**
	(0.059105)	(0.063153)	(0.057073)	(0.064485)
TFDIHC		−43.22695		
		(28.02559)		
TFDITR			28.25493*	
			(11.34589)	
TFDIFD				−54.98739*
				(19.32944)
Constant	6.675764	4.598323	8.171402*	5.469761
	(3.138726)	(3.132225)	(3.662390)	(2.985684)

Heteroskedasticity consistent standard error in parentheses, ** denotes significance level at 0.01, * denotes significance level at 0.05.

2000 to 2019, which means that the timeline is 20 years. There is no missing data. The most consistent results can be seen from two variables, which are trade openness (TR) and financial development (FD). In the case of trade openness, none of the results show any significant results. It means that trade openness is not a growth determinant in the case of Indonesia. Although, many literatures suggest otherwise, there are also those in support of this insignificant result (Brunner 2003; Carkovic & Levine 2002; Rodriguez & Rodrik 2000). On the other hand, financial development shows significant results in all four regressions: highly significant at 0.01 in the first and last equations while significant at 0.05 in the second and third equations. It shows that financial development is an important growth determinant for Indonesia. This result receives support from previous literature (Estrada et al. 2010; Guru & Yadav 2019; Khetsi & Mongale 2015).

The main independent variable in this research is TFDI. The results show that it behaves similar to human capital. It is significant in three out of four regressions. However, there is one striking difference. In the case of human capital, all of the coefficients are negative. This is interesting since human capital is expected to have positive effect on economic growth. However, it seems to not be the case in Indonesia. A similar finding was suggested by previous literature (Čadil et al. 2014; Meivitawanli 2021). One possible explanation is the use of indicator. Percentage of the population who are in school at the age of 16 to 18 is rather quantitative. The quality of education cannot be determined solely based on its length. If the teachers are not qualified, facilities are poor and contents of the subjects are not appropriate, the length of education cannot be used to measure the country's level of human capital. Therefore, the common indicator used to measure human capital should be shifted towards quality based instead of quantity based.

TFDI shows the most interesting results out of all. Not only does its significance change with the addition of interaction terms, the nature of the relationship also changes from negative to positive. In the first equation where no interaction term was added, it shows significant negative effect on economic growth. This result in itself is already interesting since it means that TFDI actually hurts host country's economy. Its negative effect is highly significant at 0.01 level of confidence. This is very important since it means that by attracting more TFDI, economic growth is slowing down. A plausible explanation is that the harms that TFDI brings with it are larger than the benefits. Examples of disadvantages that TFDI can cause include environmental damage and hindering local businesses. Environmental damage is especially harmful for tourism industry since nature and scenery are important attractions for local and foreign tourists. Hindering local businesses in tourism industry is also detrimental. Foreign companies with large capital and excellent managerial skills force local businesses to go out of business since they are unable to compete with multinationals. The tourism industry encompasses many business lines such as culinary, transportation, accommodation, souvenirs and many more. If foreign companies do not embrace local businesses, domestic enterprises will shut down one after another that will create a multiplier effect to other businesses related to it and harm the industry.

The same significant negative effect can be seen from the result of the third regression, while the effect is insignificant in the second regression. Intriguingly, in the last regression when TFDI interaction term with financial development is included, the significant effect becomes positive. It shows that financial market development is the right absorptive capacity for Indonesia. In order to absorb the benefits of TFDI, Indonesia should develop its financial market. Unfortunately, at the same time the interaction term shows significant negative effect. Looking at the coefficient, the positive effect from TFDI is 17 times higher than the negative effect of the interaction term, thus it is better to have a well-developed financial market. Financial market development as an absorptive capacity has received support from previous literature (Alfaro et al. 2004, 2010; Azman-Saini et al. 2010). The interaction term between TFDI and trade openness also shows a significant positive effect, however in equation 11.3, the negative effect of TFDI increased tremendously, showing that trade openness is not the right absorptive capacity for Indonesia.

Conclusion, Recommendation and Limitation

Tourism in Indonesia is a promising sector which receives support and attention from the government. Beautiful nature, rich culture, relaxing spas and tasty local food are only parts of what Indonesia has to offer to tourists. In order to develop its tourism industry further, the government highly encourages TFDI to build infrastructure, increase employment rate and transfer knowledge. However, empirical research on the nexus between TFDI and economic growth is not easily found especially in the case of Indonesia. This research gap is important to be filled since theories can be misleading without support from actual data. Therefore, this research examines the effect of TFDI on economic growth with the inclusion of three absorptive capacities, which are human capital, trade openness and financial market development. The results show that TFDI has a significant negative effect on economic growth unless its interaction term with financial market development is introduced into the equation. It shows that it is dangerous for the government to simply encourage TFDI without paying close attention to the development of the local financial market. The effect of TFDI on economic growth changes from significant negative to significant positive when the interaction term of TFDI and financial market development is included. This is an interesting finding since absorptive capacities have been analysed in the case of aggregate FDI but rarely used in sectoral FDI. This research found that it is applicable for TFDI in Indonesia.

Based on the results of this finding, several recommendations are made both for academicians as well as practitioners. The Indonesian government should conduct empirical research based on data before forming policies since common beliefs and notions do not always apply to every country in the world. No matter how similar countries are, there are still country specific characteristics that can render conventional ideas unsuitable. Furthermore, the government should not only focus on attracting foreign investors but also work on improving local conditions first to make sure that benefits from foreign investors can be felt by the local economy. This is a serious concern since it is not only actualising positive effect but changing negative effect to positive effect, which means that without proper internal conditions, TFDI can damage the economy. Future research can take on more absorptive capacities in the nexus of sectoral FDI and economic growth as well as utilise more indicators to represent variables such as human capital to make sure that it is well represented. Surely there are several limitations of this research, including limited data; therefore, the time span can only cover as much. There are many other growth determinants that are not included in this research. The research methodology employed is also limited; therefore, future research can adopt a more advanced methodology covering more periods and more variables.

References

Akhlas, AW 2020, 'Indonesia welcomes 16.1m foreign tourists in 2019, coronavirus clouds tourism outlook", Jakarta: *Jakarta Post,* < https://www.thejakartapost.com/news/2020/02/03/indonesia-welcomes-16-1m-foreign-tourists-in-2019-coronavirus-clouds-tourism-outlook.html>.

Al hallaq, SS, Athamneh, AB & Suleiman, HM 2019, 'The impact of foreign direct investment on the growth of the tourism sector in Jordan (1980–2016)', *Journal of Public Affairs*, vol. 20, no. 2, DOI 10.1002/pa.2005.

Alfaro, L, Chanda, A, Kalemli-Ozcan, S & Sayek, S 2004, 'FDI and economic growth: The role of local financial markets', *Journal of International Economics*, vol. 64, no. 1, pp. 89–112, DOI 10.1016/S0022-1996(03)00081-3.

Alfaro, L, Chanda, A, Kalemli-Ozcan, S & Sayek, S 2010, 'Does foreign direct investment promote growth? Exploring the role of financial markets on linkages', *Journal of Development Economics*, vol. 91, pp. 242–256, DOI 10.1016/j.jdeveco.2009.09.004.

Andrew, S 2020, *Scientists just found one of the world's largest flowers blooming in an Indonesian jungle*, CNN Travel, https://edition.cnn.com/2020/01/03/world/worlds-largest-flower-indonesia-scn-trnd/index.html.

Apuke, OD 2017, 'Quantitative research methods: A synopsis approach', *Arabian Journal of Business and Management Review*, vol. 6, pp. 40–47, DOI 10.12816/0040336.

Azman-Saini, WNW, Law, SH & Ahmad, AH 2010, 'FDI and economic growth: New evidence on the role of financial markets', *Economics Letters*, vol. 107, no. 2, pp. 211–213, DOI 10.1016/j.econlet.2010.01.027.

Balasubramanyam, VN, Salisu, M & Sapsford, 1996, 'Foreign direct investment and growth in EP and IS countries', *Economic Journal*, vol. 106, no. 434, pp. 92–105, DOI 10.2307/2234933.

Barrowclough, D 2007, 'Foreign investment in tourism and small island developing states', *Tourism Economics*, vol. 13, no. 4, pp. 615–638, DOI 10.5367/000000007782696122.

Batten, JA & Vo, XV 2009, 'An analysis of the relationship between foreign direct investment and economic growth', *Applied Economics*, vol. 41, no. 13, pp. 1621–1641, DOI 10.1080/00036840701493758.

Borensztein, E, De Gregorio, J & Lee, JW 1998, 'How does foreign direct investment affect economic growth?', *Journal of International Economics*, vol. 45, no. 1, pp. 115–135, DOI 10.1016/S0022-1996(97)00033-0.

Brunner, AD 2003, 'The long-run effects of trade on income and income growth', *IMF Working Papers*, vol. 03, no. 37.

Butler, RA 2016, The top 10 most biodiverse countries, *Mongabay*, https://news.mongabay.com/2016/05/top-10-biodiverse-countries/.

Čadil, J, Petkovová, L & Blatná, D 2014, 'Human capital, economic structure and growth', *Procedia Economics and Finance*, vol. 12, pp. 85–92, DOI 10.1016/S2212-5671(14)00323-2.

Calvo, MB & Sanchez-Robles, B 2003, 'Foreign direct investment, economic freedom and growth: New evidence from Latin America', *European Journal of Political Economy*, vol. 19, no. 3, pp. 529–545.

Carkovic, M & Levine, RE 2002, 'Does foreign direct investment accelerate economic growth?', *University of Minnesota Department of Finance Working Paper.*

Cheung, T 2017, Your pick: World's 50 best foods, *CNN Travel*, https://edition.cnn.com/travel/article/world-best-foods-readers-choice/index.html.

CNN Indonesia 2017, *Spa Nusantara Festival 2017 Pamerkan Layanan Spa Terbaik*, https://www.cnnindonesia.com/gaya-hidup/20170714135400-307-227915/spa-nusantara-festival-2017-pamerkan-layanan-spa-terbaik.

Conservation International n.d., *Bird's Head Seascape*, https://www.conservation.org/places/birds-head-seascape.

Enjoy Jakarta n.d., *Istiqlal Mosque*, Department of Tourism and Culture of Jakarta, http://jakarta-tourism.go.id/visit/blog/2020/06/istiqlal-mosque

Estrada, G, Park, D & Ramayandi, A 2010, 'Financial development and economic growth in developing Asia', *ADB Economics Working Paper Series*, vol. 233, DOI 10.2139/ssrn.1751833.

Fauzel, S 2021, 'FDI and tourism futures: a dynamic investigation for a panel of small island economies' *Journal of Tourism Futures*, vol. 7, no. 1, pp. 98–110, DOI 10.1108/JTF-05-2018-0026.

Fauzel, S, Seetanah, B & Sannassee, RV 2017, 'Analysing the impact of tourism foreign direct investment on economic growth: Evidence from a small island developing state', *Tourism Economics*, vol. 23, no. 5, pp. 1042–1055, DOI 10.1177/1354816616664249.

Garson, DJ 2012, *Testing Statistical Assumptions*, Statistical Associates Publishing, North Carolina.

Gbakou, M, Jallab, MS & Sandretto, R 2009, 'Foreign direct investment, macroeconomic instability and economic growth in MENA countries', *GATE Working Paper*, pp. 08–17.

Ghofar, M 2015, Kemendikbud: Wajib Belajar 12 Tahun dimulai 2016, *Antara News*, https://www.antaranews.com/berita/503076/kemendikbud-wajib-belajar-12-tahun-dimulai-2016.

Guinness World Records, 2021, *Largest Buddhist Temple*, https://www.guinnessworldrecords.com/world-records/largest-buddhist-temple/.

Guru, BK & Yadav, IS 2019, 'Financial development and economic growth: panel evidence from BRICS', *Journal of Economics, Finance and Administrative Science*, vol. 24, no. 47, pp. 113–126, DOI 10.1108/JEFAS-12-2017-0125.

Hananto, A 2020, *Lost World of Asia, These Rare Animals Can Only be Found in Indonesia*, Seasia, https://seasia.co/2020/10/29/lost-world-of-asia-these-rare-animals-can-only-be-found-in-indonesia.

Iamsiraroj, S & Ulubaşoğlu, MA 2015, Foreign direct investment and economic growth: A real relationship or wishful thinking?', *Economic Modelling*, vol. 51, pp. 200–213, DOI 10.1016/j.econmod.2015.08.009.

Iamsiraroj, S 2016, 'The foreign direct investment-economic growth nexus', *International Review of Economics & Finance*, vol. 42, DOI 10.1016/j.iref.2015.10.044.

Indonesia Investment Coordinating Board 2018, *Kepala BKPM: Investor Minati Sektor Pariwisata RI*, Ministry of Investment Republic of Indonesia, https://www.bkpm.go.id/id/publikasi/detail/berita/kepala-bkpm-investor-minati-sektor-pariwisata-ri.

Invest Islands 2019, *The "10 New Bali" Project In Indonesia: What Is It, And Why Will It Shape The Future Of Southeast Asia's Tourism Industry?*, Invest Indonesian Islands, https://invest-islands.com/ten-new-bali-project/.

IPB University 2016, *Many Local Fruits Exhibited in the Indonesia Fruit of 2016*, https://www.ipb.ac.id/news/index/2016/11/many-local-fruits-exhibited-in-the-indonesia-fruit-of-2016/2e02946cb308b0f715ecf21fd0a09a21#:~:text=Minister%20of%20Agriculture%2C%20Amran%20Sulaiman,and%20watermelon%2C%20%22said%20Amran.

Khaliq, A & Noy, I 2007, 'Foreign direct investment and economic growth: Empirical evidence from sectoral data in Indonesia', *Working Papers 200726, University of Hawaii at Manoa, Department of Economics*.

Khan, MA 2007, 'Foreign direct investment and economic growth: The role of domestic financial sector', *PIDE Working Papers*.

Khetsi, QS & Mongale, IP 2015, 'The impact of capital markets on the economic growth in South Africa', *Journal of Governance and Regulation*, vol. 4, no. 1, pp. 154–163, DOI 10.22495/jgr_v4_i1_c1_p6.

Kholdy, S & Sohrabian, A 2005, 'Financial markets, FDI, and economic growth: Granger causality tests in panel data model', *EFA 2005 Moscow Meetings*, DOI 10.2139/ssrn.676085.

Kumairoh 2019, *Kemenpar Beberkan 10 Program Pariwisata RI yang Pace Pendapatan Devisa*, Warta Ekonomi, https://www.wartaekonomi.co.id/read245267/kemenpar-beberkan-10-program-pariwisata-ri-yang-pacu-pendapatan-devisa.

Li, X & Liu, X 2005, 'Foreign Direct Investment and economic growth: An increasingly endogenous relationship', *World Development*, vol. 33, no. 3, pp. 393–407, DOI 10.1016/j.worlddev.2004.11.001.

Lindblad, JT 2015, 'Foreign direct investment in Indonesia: Fifty years of discourse', *Bulletin of Indonesian Economic Studies*, vol. 51, no. 2, pp. 217–237, DOI 10.1080/00074918.2015.1061913.

Meivitawanli, B 2018, 'Foreign direct investment in tourism and economic growth: Panel data of OECD countries', *International Journal of Advanced Engineering Research and Science*, vol. 5, no. 9, pp. 287–295, DOI 10.22161/ijaers.5.9.34.

Meivitawanli, B 2021, 'Research on the relationship between provincial foreign direct investment and economic growth: An empirical study in Indonesia', *Journal of International Studies*, vol. 14, no. 1, pp. 241–257, DOI 10.14254/2071-8330.2021/14-1/17.

Menyari, YE 2020, 'Effect of tourism FDI and international tourism to the economic growth in Morocco: Evidence from ARDL bound testing approach', *Journal of Policy Research in Tourism, Leisure and Events*, DOI 10.1080/19407963.2020.1771567.

Montgomery, DC, Peck, EA & Vining, GG 2021, *Introduction to Linear Regression Analysis*, New Jersey: John Wiley & Sons.

Organisation for Economic Co-operation and Development 2018, *OECD Benchmark Definition of Foreign Direct Investment*, https://www.oecd.org/daf/inv/investmentstatisticsandanalysis/40193734.pdf.

Organisation for Economic Co-operation and Development 2020, *Indonesia*, OECD iLibrary, viewed 21 May 2021, https://www.oecd-ilibrary.org/sites/6c4fc1cd-en/index.html?itemId=/content/component/6c4fc1cd-en.

Pratoomchat, P 2017, 'Tourism-led growth hypothesis and foreign direct investment in ASEAN in tourism and opportunities for economic development in Asia', *Working paper for the 55th Annual Meeting of the Southeast Conference Association for Asian Studies (SECAAS), James Madison University*, DOI 10.13140/RG.2.1.3622.2804.

Rashid, IMA, Samah, IHU & Amlus, MH 2020, 'Simultaneous equation model of FDI in tourism, tourism growth and poverty: development of model specification', *International Journal of Scientific & Technology Research*, vol. 9, no. 4, pp. 3610–3613.

Rodriguez, F & Rodrik, D 2000, 'Trade policy and economic growth: A skeptic's guide to the cross-national evidence', In Bernanke, BS & Rogoff, K (eds.), *NBER Macroeconomics Annual 2000*, Cambridge: National Bureau of Economic Research, pp. 261–338.

Samimi, AJ, Sadeghi, S & Sadeghi, S 2013, 'The relationship between foreign direct investment and tourism development: evidence from developing countries', *Institutions and Economies*, vol. 5, no. 2, pp. 59–68.

Steiner, C 2010, 'An overestimated relationship? Violent political unrest and tourism foreign direct investment in the middle east', *International Journal of Tourism Research*, vol. 12, no. 6, pp. 726–738, DOI 10.1002/jtr.788.

The World Tourism Organization 2020, *Indonesia*, United Nations, https://webunwto.s3.eu-west-1.amazonaws.com/s3fs-public/2020-10/indonesia.pdf.

Warta Ekspor 2018, *Mie Instan Sangat Disukai Orang Afrika (PEN/MJL/009/08/2018)*, Ministry of Trade, Republic of Indonesia, http://djpen.kemendag.go.id/app_frontend/admin/docs/publication/5361548127292.pdf.

Widiatedja, IGNP 2019, 'FDI on tourism and the environmental damage: what Indonesia's trade and investment agreements can do?', *Yuridika*, vol. 34, no. 2, pp. 303–324, DOI 10.20473/ydk.v34i2.7927.

World Bank 2021, *The World Bank in Indonesia,* < https://www.worldbank.org/en/country/indonesia/overview#1>.

World Economic Forum 2019, *The Travel & Touism Competitiveness Report 2019*, http://www3.weforum.org/docs/WEF_TTCR_2019.pdf.

World Factbook n.d., *Indonesia*, Central Intelligence Agency, viewed 20 May 2021, Washington, DC, https://www.cia.gov/the-world-factbook/countries/indonesia/.

World Heritage Center n.d., *Prambanan Temple Compounds*, United Nations Educational, Scientific and Cultural Organization, Paris, https://whc.unesco.org/en/list/642/.

World Travel & Tourism Council 2019, *Key Highlights – Economic Impact of Cities 2019*, London, UK, https://wttc.org/Portals/0/Documents/Reports/2019/City%20Travel%20and%20Tourism%20Impact%20Key%20Highlights%20Dec%202019.pdf?ver=2021-02-25-201317-800.

World Wild Fund For Nature n.d., *The Richest Coral Reef on Earth & Indonesia's Top Liveaboard Diving Destination*, Gland, Switzerland, https://wwf.panda.org/discover/knowledge_hub/where_we_work/coraltriangle/coraltrianglefacts/places/rajaampatindonesia/?.

Yosephine, L 2018, *Jakpost guide to Taman Mini Indonesia Indah*, Jakarta: Travel Guide by The Jakarta Post, https://www.thejakartapost.com/travel/2018/09/08/jakpost-guide-to-taman-mini-indonesia-indah.html#:~:text=Taman%20Mini%20Indonesia%20Indah%20(TMII,the%20archipelago's%20rich%20cultural%20heritage.

Yuniarni, S 2016, *Unity in Diversity: Indonesia's Six Largest Ethnic Groups*, Jakarta: Jakarta Globe, https://jakartaglobe.id/culture/unity-diversity-indonesias-six-largest-ethnic-groups/.

Zekarias, SM 2016, 'The impact of foreign direct investment (FDI) on economic growth in eastern Africa: Evidence from panel data analysis', *Applied Economics and Finance*, vol. 3, no. 1, pp. 145–160, DOI 10.11114/aef.v3i1.1317.

12 Experienced 'Realities' of Tourism Foreign Direct Investment

Empirical Evidence from Barbados

H. Cristina Jönsson

Introduction

There remain notably few studies on tourism foreign direct investment (TFDI) and where research exists, the focus has been mainly on the overall structure of TFDI (see for example Gupta, 2015) and the political economy of TFDI (see for example Bianchi, 2018). Broad scholarly attention has been paid to the socio-economic consequences of tourism on local peoples' lives, however little empirical evidence exists on the importance of local impacts of TFDI. A notable exception is the work of UNCTAD (2007), arguing a need for tourism "... to be managed carefully if it is to yield the desired benefits without undermining the local economy and the environment, social traditions and cultural resources". In the context of tourism-dependent, small island developing states (SIDS), Scheyvens and Momsen (2008 as cited in Alam and Paramati, 2016: 114) have argued that "although tourism has a significant contribution to economic development, it does not have a considerable influence on poverty alleviation...In fact, in some [small island developing] states, tourism increases existing income inequalities". To date, however, it is difficult to locate academic research that specifically focuses on the experienced local impacts of TFDI in SIDS, highlighting a discernible research gap to which this chapter intends to contribute.

By analysing facts, and the interpretation of facts, by various respondents, this research has the potential to contribute to the academic understanding of hotel employees' experienced realities of TFDI impacts in a small island tourism economy. A subjective ontology to "reality" is adopted. Reality exists but is experienced and open to subjective interpretation rather than fixed (epistemological realism). It is conceived that TFDI exists but the nature of its impacts are open to multiple perceptions or cognitive subjective interpretations, in line with a realist epistemology. The nature of being (ontology) is viewed as subjective with reality being open to multiple views or perceptions, shaped by the personal experiences of individuals. This research takes a critical realist ontological stance, which reflects the author's view of reality; that there is a mind-independent external reality and that it can be known or accessed but this happens cognitively by individuals.

DOI: 10.4324/9781003155492-15

Critical realism denies that there is any certain knowledge of the world, and accepts the possibility of alternative valid accounts of any phenomenon. In other words, there are different valid perspectives on reality. In his classical statement, Sapir (1958) argued that "the worlds that different societies live in are different worlds, not simply the same world with different labels attached" (1958: 69). The perceptions of hotel employees in Barbados will be presented as their subjective, experienced "realities".

Barbados is a mature Caribbean tourist destination and it is one of the leading island destinations in the Southern Caribbean for British and North American tourists (Jönsson, 2005). Barbados has very few natural resources; therefore, tourism continues to receive increased attention as an important sector of Barbados' economy. In 2018, Barbados received 680,000 over-night visitors (The World Bank, 2018), mainly from the UK, USA, and Canada with the UK market being the dominant tourist market of arrivals, followed by the USA. Based on available data, tourism generated 50% of Barbados' foreign exchange earnings (BTI, 2015) and accounts for 12.9% of GDP (WTTC, 2017).

International hotel chains and tour operators have shown interest in Barbados as a tourism destination since the 1960s. With an increase in visitor demand, the Barbados government saw the potential of the tourism industry to diversify its economy. There has been significant FDI in Barbados, mainly in the development and sales of real estate, hotels, and condominiums. FDI peaked in 2007 with $675.6 million, of which real estate transactions accounted for over 60% (Central Bank of Barbados, 2010). Due to external shocks such as the global economic downturn, FDI in Barbados declined to from a pre-crisis average of USD $403 million in 2005–2007 to USD $254 million in 2015 (UNCTAD, 2016). Barbados' "plantocracy" (the Europeans who owned most of the land and slaves in West Indian Colonies) is mainly involved in tourism indirectly through ownership of land, commerce, seaport, tax-free outlets, and construction companies. The Barbados government is officially committed to ensuring that tourism will improve the life and economic development of the people of Barbados (Barbados Tourism Master Plan, 2014) and has a comprehensive social development policy emphasising human capital development. In recognising tourism's actual and potential contribution to the economy, the Barbados government is giving the sector a higher political priority than before. Attempts at tapping into new tourism markets such as Brazil and China have been made (Jönsson & Harrison 2015). Additionally, the tradition sun, sea, and sand tourism product is being diversified, for example, by including sports tourism (Jönsson, 2008). As a result, economic policy objectives increasingly centre on tourism's ability to generate employment, earn foreign exchange, and contribute significantly to government revenue.

Barbados has an extensive history as it relates to FDI and tourism FDI in the context of the Caribbean. It is partly due to FDI that Barbados has been economically successful over the years. FDI has been able to provide and sustain employment in Barbados while facilitating technological growth. Yet, as simple as it may seem, it is a complex topic that depends on a multiplicity of factors for it to be successful. Factors ranging from policy making, infrastructure, accessibility, and stakeholder collaboration to ensure that the economic and social environment is conducive for development and FDI.

Literature Review

Trans National Corporations and FDI

A long history of arguments in favour of and against FDI may be recognised; it has been purported that FDI may lead to economic growth and productivity increases as a whole but the risks associated with FDI have been emphasised. Bayulgen (2010) explores these opposing views in two alternative conceptualisations of FDI impacts. The first model, "the benign model", is eminently positive and emphasises the net addition of inputs that foreign investors may bring to a domestic setting of dynamic competition. It describes how FDI can be used to interrupt the forces which drive under-development in a country. As explained by Bayulgen (2010, p.72) it "assumes an inherent link between increased economic inter-dependence, socio-economic development, and democracy". This modernist approach essentially argues that the presence of Trans National Corporations (TNCs) should result in an increase in efficiency, expanded output, and higher economic growth in the host country. In line with these potential benefits, TNCs have been identified one of the primary international forces of economic, social, and environmental change in the global economy. Furthermore, Bury (2001) has claimed that TNCs affect not only economic development, technology, skills, trade, and employment, but also the living conditions and natural environment of billions of people.

Bayulgen's second model, "the malign model", is much more problematic and distinctly negative. It emphasises the potentially distortionary impact that foreign investors from competitive international industries may have on domestic economies that are themselves challenged by market imperfections, such as their heavy reliance on imports to drive socio-economic growth. In such cases, FDI can also have a noticeably negative impact on the host country's prospects for development. This alternative model of FDI and development emphasises the potential malign interaction between TNCs and host economies (Bayulgen, 2010). It reflects much of the criticism of TNCs whereby there is a possibility that FDI leads to government-related distortions by obstructing price and profit regulations, and antitrust laws. In such cases, Viscusi et al. (2005) argue that this gives governments explicit ability to deny companies from merging when that merger is perceived as threatening the competitive structure of the market. Other prohibitive actions that can arise from the interaction between TNCs and host economies include the introduction of policies regulating the migration of labour, wages, and the obstruction of laws designed to prevent socially undesirable practices, like income inequality regulations. In particular, there is scope for potential negative interaction to occur between TNCs in imperfectly competitive industries and host economies with imperfectly competitive domestic markets and some researchers have suggested that FDI is a product of market imperfections. Hymer (1976), for example, has articulated his view that the emergence of TNCs is a response to an imperfect global market environment. This, he claims, was either occurring naturally or imposed by government through imposition of tariffs and foreign exchange control to protect the domestic industry.

Buckley and Casson (1976) support an "internalisation theory". This theory postulates that for a firm to engage in FDI it must have some advantage over rival firms in other national markets. This advantage can be in the form of superior technology, management know-how, or marketing and distribution skills and is characterised by enabling a TNC to compete successfully with domestic competitors. It is notable that these factors of advantage, when capitalised upon by the TNCs, are often rebuffed by local competitors. However, it is also noteworthy that in some instances, the advantages of, and access to resources by the TNCs can assist and improve the product offerings of local competitors. A very apt example of this somewhat ironic occurrence exists in Barbados. One tourism TNC, Sandals Resorts International (SRI), purchased a property near a popular tourism belt, St. Lawrence Gap, in Barbados. SRI embarked on a shoreline stabilisation project to construct groynes to prevent coastal erosion and replenished sand on the beach for its resort expansion (Bradshaw, 2015). Since all beaches in Barbados are public beaches, guests from surrounding properties and local vendors plying their trade on the beach are expected to benefit from the investment by SRI to upgrade the beach.

In order to explain the behaviour of TNCs and their motivations for engaging in FDI, Dunning (1979) developed "the eclectic paradigm of international production", also referred to as OLI based on the three decision factors it proposes: ownership, location, and internalisation. It is a holistic, yet context-specific, framework for analysing FDI determinants and it combines several isolated theories of international economics in one approach. The framework notably distinguishes between different forms of international activities: exporting, FDI, and licensing. The OLI-factors encompass both firm-specific (ownership) and country-specific (location) motives for international activities as well as highlighting internalisation advantages.

From the firms' perspective, ownership and internationalisation determine whether or not FDI is an attractive strategy. The ownership-specific advantages include unique property rights, and intangible aspects such as product innovation. Dunning (2001: 175) sees these advantages as "any kind of income generating asset that allow firms to engage in foreign production". TNCs' country selection for FDI is influenced by locational advantages, which relate principally to tariff barriers, labour costs, and the presence of competitors. SRI in Barbados provides a somewhat subtle example of a TNC capitalising on such benefits. As Wreath (2015, para 1, 2.12) reports "Barbados is a destination beloved of celebrities, jetsetters and anyone seeking an elegant retreat...Sandals Barbados is the island's only 5-star Luxury All-inclusive resort designed exclusively for couples." This statement does well to showcase how Sandals capitalised on Barbados' positioning as a luxury high-end destination and essentially crafted a niche for itself where, in effect, it has a monopoly existence in its class. Dunning (2001: 177) further suggests that internationalisation "relates to the costs and benefits of different modalities of coordinating multiple economic activities". While the OLI framework helps to evaluate the motives for a TNC to engage in FDI, greater insight is needed as to how the impact of TNCs in a specific country is perceived by local people.

Tourism Foreign Direct Investment

FDI, in general, has been praised for being able to help developing countries through the creation of employment and the supply of capital and resources such as managerial skills, technological and marketing and promotion assets. For many developing countries, FDI has surpassed official development assistance (ODA), remittances, and portfolio investment flows, and become the largest source of external finance (World Bank, 2017). On the other hand, FDI has also received criticism, not only for economic "leakages" (Pratt, 2015) but also for the possibility that it obstructs policies regulating the migration of labour, wage policies, and other laws that hinder socially undesirable practices (Kaulihowa & Adjasi, 2018). Hence, in tourism-dependent countries, efforts need to be directed towards the creation of linkages that can lead to benefits for local people through TFDI. Too much FDI may lead to excessive trust in FDI which in turn may lead to high risk for the host community. In a Caribbean context, the WTTC (2002) has outlined the case for paying attention to local tourism impacts:

> There is fairly widespread concern that the benefits from tourism in the Caribbean might not be reaching those who need them most and that, despite the impressive growth statistics, tourism is not doing as much as it could to relieve existing problems of unemployment, poverty and social dislocation.

In tourism-dependent countries, tourism is often reported to be seen by governments as a solution for the country's economic ills. As a result the tourism sector has been increasingly prioritised as a potential path for economic and human development in less economically developed countries, or LEDCs (Khan et al., 2020). FDI has become a key component of national development strategies for many countries around the world. However, there are concerns about the power that TNCs wield over destinations

(Meyer, 2013) and the fear is that due to so-called "leakages" the share of local gains from tourism is often small (Chirenje et al., 2013) Indeed, as Meyer (2006: 4) has noted, "The potential of tourism as a local development option has been criticised due to persistent claims that the industry is plagued by high import content and subsequently high leakages". According to UNCTAD (2020), FDI is growing fastest in LEDCs, and new opportunities for socio-economic development are being offered.

According to UNCTAD (2010), SIDS should be recognised to be a distinct group of developing countries facing specific social, economic, and environmental vulnerabilities. More specifically, they face issues of a small domestic market, lack of scale, and high transport costs (Pratt, 2015). This has ultimately led to the end of the export industries, forcing these states to rely on their natural resources of sun, sea, and sand. In terms of level of development, there is however potential for SIDS to grow their share of tourism revenue. Between 1990 and 2016 developing countries as tourism destinations increased their share of international tourism receipts from 18% to 40% (UNWTO, 2017). Furthermore, the United Nations World Tourism Organisation (UNWTO) predicted that by 2020 there would be 1.6 billion tourists; up from 0.8 billion in 2006 and, based on the pre-pandemic rate of growth, the majority of these would be visiting developing market destinations. However, 2020 recorded a 73% fall in worldwide international tourist arrivals, making it the worst year on record for the tourism sector (UNWTO, 2021a, 2021b). With a focus on Barbados, tourism is credited with helping to diversify Barbados' economy away from traditional commodity-based activities (mainly sugar) and the country has enjoyed rapid tourism growth, adding weight to its recognition from government as a key economic sector (Barbados Ministry of Finance and Economic Affairs, 2011). However, at the time of writing the impact of the pandemic has had ripple effects across the economy and the recovery prospects of tourism are uncertain.

There is considerable literature on the impact of FDI and growth generally but there are a limited number of studies on FDI within the field of tourism that focus on the context of SIDS. The attention of these studies has been mainly on the overall structure of TFDI (see, for example, (Fauzel, 2020; Vítová et al., 2019) where available FDI stock data has been utilised by calculating FDI flows over a period of time to analyse and reveal the pattern and scale of FDI in tourism. This type of analysis can help inform the role and significance of TFDI in countries where data is available. Yet, whilst detailed country FDI statistics are available from more developed countries, there remain challenges to presenting an accurate picture in countries with little or no available data, such as is the case with Barbados. This is another reason why a study relating to TFDI in Barbados needs to essentially include qualitative data.

Local impacts of tourism and FDI are perhaps different between SIDS and other geographical locations given the economic challenges, remoteness, and livelihood trajectories of the destination areas. However, although local-level perceptions of tourism have been widely examined in a range of peripheral contexts (including, Sharpley, 2014; Almeida-García et al., 2016), few studies have been carried out in SIDS (see for example, Nunkoo & Ramkissoon, 2011; Simão & Môsso, 2013). This is despite emphasis placed on the importance of tourism to SIDS. Furthermore, although research has shown that TFDI is beneficial to tourism-dependent nations, few studies have focused on local interpretations or "experienced realities" of TFDI impacts. A qualitative analysis of the local impacts of TFDI is more appealing and practical than taking an econometric approach since the central question of this chapter is: "what are the 'realities' of TFDI for local hotel employees in Barbados?" This study aims to contribute to that gap in knowledge.

UNCTAD (2007) has argued that more empirical research on TFDI is needed for developing countries to make the most out of the sustainable benefits offered by TFDI and to gain awareness of appropriate policies that can be applied to maximise local gains from TFDI. The research in this chapter aims to contribute to this research gap by emphasising the voices of hotel employees in an investigation of impacts of TFDI in Barbados. There remains a lack of consensus in the existing literature on the impacts that FDI has on local people and there is a scarcity of research that considers local perspectives on the outcomes of TFDI in a tourism-dependent economy.

Data Collection

Focus group interviews were selected as an appropriate method for accessing a broad spectrum of resident perceptions of TFDI impacts. The nature of tourism activity in Barbados had to be acknowledged as part of practical data collection considerations. As an SIDS with a population of approximately 286,000 and with tourism being the primary economic activity, it was recognised that all Barbadian residents would have some level of interaction with the tourism sector. As a small destination, only 166 square miles in size, tourism activity is not generally confined to enclaves or traditional tourist spaces in the same way that the literature ascribes these attributes to larger and more developed destinations (Schulte & Dissart, 2015). While tourism activity in Barbados is to some extent concentrated in certain areas (such as the West and South coast), these locations locally known as "tourist belts" are not tourist enclaves in the traditional sense as they are equally accessible to, utilised by, and inhabited by the local population. In essence therefore, no segment of the Barbadian population can truly be isolated from the implications of tourism activity, especially as a result of the island's extreme reliance on tourism for economic growth and development. This was important to recognise in terms of the design of data collection. It held implications for the design of the focus group interview questions and also the sampling strategy.

Focus groups were used due to the ease of acquiring information. Culturally, Barbadians enjoy group discussions. Conversely, they are reluctant to one-on-one conversations where they feel that they are giving information, such as an interview. In a small island society such as Barbados there is a fear of being victimised, which is one of the reasons that semi-structured interviews are more challenging. In the focus group interviews their participation was less formal and the participants got carried away discussing the topic.

Discussion

The idea that there exist multiple realities and lived experiences underpins the ethos of this chapter. An interpretive approach is adopted and the purpose is to capture hotel employees' "realities" of local impacts of TFDI in Barbados. In order to extend current TFDI knowledge, this chapter uses qualitative data as opposed to applying the research methods generally adopted in TFDI studies, such as econometric analysis (see for example Mustafa, 2019) and statistical analysis (see for example Arain et al., 2020). The use of focus group interviews with local hotel employees allow for a broad spread of views to be captured whilst placing emphasis on personal experiences, interpretations, and views on the consequences of tourism development through FDI.

This part of the chapter presents the results of the primary research on local resident perceptions of TFDI impacts in Barbados, focusing on TFDI (hotel) employees. These results are based on thematic analysis of focus group interview data. "Realities" of local TFDI impacts are identified and these are critically evaluated to identify relationships between these dimensions and currently known impacts of TFDI.

The results of the focus group interviews with residents who are TFDI employees are presented. Their experienced "realities", the sub-themes, and themes of the data are then considered before a summary of results is offered. The interviewees in the hotel employee focus group interviews were self-selected. There was a balanced mix of gender and age. Interviewees had been employed in hotels between 1 and 20 years.

In total, 59 TFDI employees of hotels in Barbados were interviewed from six different hotels. The findings are discussed for employees by each hotel in turn before considering an overall hotel employee "reality". The findings are made anonymous and the hotels have been coded. For each hotel the employees have been allocated a number (Employee 1, Employee 2, Employee 3, and so on). The following discussion of the research results follows the themes of the hotel employee data in Table 12.1.

The 'Experienced Reality' of Hotel Employees

Among the TFDI employees participating in the study, there was acknowledgement of the high contribution of tourism to the island of Barbados. As expected, positive employment impacts were noted, often relating to their own livelihoods, alongside macro-economic impacts.

For the hotel employees interviewed, a number of shared salient sub-themes and themes were identifiable in their experiences or "realities" of the local impacts of TFDI on tourism development (Table 12.1).

Table 12.1 Sub-themes and themes of the hotel employee data

Sub-themes	*Themes*
Exposure to global cultures	• Employment and skills development
Access to international training	
• Contribution to tourism in Barbados	• Role of foreign direct investors in national development
Employment creation	
• TFDI contributes to poverty alleviation through the individual corporate social responsibility (CSR)	• The scale and value of TFDI
• TFDI properties impact on resource growth which domestic firms do not necessarily have access to	• Power and control
Local employees' restricted growth to executive levels in TFDI properties	
TFDI properties are less receptive of rights and powers of labour unions and labour bargaining power	

Theme 1 – Employment and Skills Development

It was notable that local resident interviewees discussed TFDI more primarily in the context of its direct impact on citizens more so than at the macro-level of the country. There was indeed a general acceptance that substantial benefits can indeed be achieved from being associated with tourism entities. Based on the benefits identified however, it is extremely noteworthy that a perception existed among local resident interviewees that some types of the individual benefits achieved are possibly restricted to those employed within foreign entities. For example, one interviewee from Hotel #1 suggested that by being employed at the hotel with its contingent of non-national staff, their exposure to global cultures expanded (Employee 1, Hotel #1). This is a particularly important point when one considers that individuals from poorer communities or those who are lesser skilled often do not have opportunities to be exposed directly to global cultures. In this case, access to opportunities goes beyond economic opportunities and extends to life chances linked to socio-cultural development.

Moreover, as explained by Employee 2, Hotel #1:

> X has provided me with employment and the chance to training overseas and even employment opportunities in the other X franchises. In the training process I am exposed to different techniques and being part of a well-known brand.

It is particularly interesting that in the response provided above, the interviewee made specific mention of the impact that the diverse training techniques in foreign brands has had on their development. It is clear therefore that the international exposure to and perspectives of international brands provide a broader scope for improving human resource capacity of citizens in the destination working with the properties. This reinforces the position of the benign model recalled by Bayulgen (2010) which suggests that FDI has a substantial impact on resource growth which domestic firms do not necessarily have access to. As deduced from the comment above by the respondent, this improvement is argued to be perpetuated by the fact that the international exposure and opportunity gives the human resources the potential to access standards with international credence, influenced by the diversity in global cultures the organisation would have to develop their standards to respond to. From a local perspective, it was also indicated that the international standards adopted by international properties force the competition to also improve on their standards because maintaining lower standards reduces the local competition's competitive advantage.

Notwithstanding these benefits however, some interviewees alluded to a perception that local citizens are mostly only offered front line positions with few opportunities to advance to middle and upper level management positions. While this is an extremely valid concern it is also worth noting that from an employee and skills development perspective, these international entities have certain service standards influenced by international cultures which necessitate having management which appreciates, and has already been indoctrinated with said culture. Recognition of the importance of standards for these organisations highlights this perception. In essence, while this offers some explanation as to why newly established FDI entities do not fill middle and upper level management positions with locals, it does not offer support to FDI entities established in the destination for prolonged periods having had the opportunity to "indoctrinate the culture into locals".

More importantly though, the restricted growth to executive levels for locals in TFDI properties does not necessarily hold true to the philosophy of Bayulgen's (2010) benign model. In spite of there being growth in the resource capacity of locals employed within TFDI properties, the reality that locals do not traditionally rise to executive positions within that destination reflects the malign perspective more (Bayulgen, 2010), given this restriction on domestic human resource growth. This reality offers some explanation as to why the work of Brohman (1996) did not find a strong enough correlation between TFDI and poverty alleviation. To further reinforce the weak potential for such a correlation is the fact that macro poverty alleviation policies are not necessarily integrated into TFDI policies. This study finds that TFDI contributes to poverty alleviation more so through the individual corporate social responsibility (CSR) efforts of the foreign investors. Since CSR is affected by variables such as management style and power relationships (Coles et al., 2013), its contribution to poverty alleviation varies.

Moreover, at the outset, it may appear that when employees use their training and background with TFDI properties to seek international jobs this has entirely negative implications because of the local loss of that human resource. However, relocating and training local human resources can be viewed positively if after receiving the foreign employment and training the individual returns to the destination to use the experiences gained. Additionally, the destination is also likely to benefit from any remittances from those individuals. Moreover, this TFDI-fuelled exodus of human resources does not have the same impact on market imperfections as the price distortions explained by Bayulgen (2010). The training and opportunities provided by TFDI properties has been argued by interviewees to force other local properties to improve their product and service offerings or risk losing market share to TFDI properties. Therefore, in spite of the fact that leakages are a consequence of TFDI, the skills development and competition in facilities may be seen to benefit the destination from a human resource perspective in two primary ways. First, competitors are forced to improve their service standards and second, local companies can benefit from the skillset that persons trained by the TFDI properties develop. These findings attest that while Bayulgen's (2010) malign model considers the implications of an exodus of resources, a TFDI-fueled exodus of human resources does not necessarily translate into permanent negative implications.

The following response adds testimony to the claim by Fauzel (2016) that while in many SIDS, openness to TFDI is high there are often limited structured policy approaches to capitalise on TFDI. An employee from one of the local TFDI properties vexingly said, "We have a shop steward here, we are members of the union but everybody here frighten, we can't go the meetings, we don't even know when the meetings are". In spite of Barbados' long-established recognition of the rights and powers of labour unions and labour bargaining power, the management of this TFDI property is less receptive to this domestic practice of collective worker's representation. This fear has the potential to undermine the entire collective bargaining process across the country and is therefore an excellent example of a potential consequence of unfettered and unstructured controls on TFDI in labour management.

In essence, the quote referenced earlier by Employee 2 of Hotel #1 reflects an ideology espoused within the study that local staff perceive that foreign employment gives them access to certain benefits they would not ordinarily receive from employment within local entities. It was not surprising therefore that Employee 5 of Hotel #1 declared "I was able to travel and be a part of a well-known brand". It is particularly interesting, however, that this employee correlated their satisfaction with their organisation being a "well-known brand" which at its core implies that the employee would express the same level of satisfaction and recognition if it were a local well-known brand. However, the preceding evidence does not largely support this postulate. This rebuttal is based on the premise that respondents made correlations between well-known brands and, (1) the international cultures within the brand and (2) the ability to travel to other branches of the brand. Certainly on that basis one could question that, if local brands were able to offer this international exposure and travel potential, would local employees regard both local and international establishments equally in terms of personal benefits? As a result of this perceived international exposure from being associated with international brands, these locals develop a sense of commitment/loyalty to foreign entities for these opportunities. For example, one respondent suggested "I was offered a career opportunity overseas but I declined." (Employee 4, Hotel #1).

Theme 2 – Role of Foreign Direct Investors in National Development

TFDI was also perceived as having a role in the development of overall national economies. There were indications that there is a dependence on the international community to provide resources to develop areas of the local tourism sector which were either too substantial or required too many resources to be developed by the local investment community. In this context, foreign investors were perceived to fill a void that could not be managed domestically. One resident interviewee who was a member of the Tourism Association discussed the very topical issue of Hotel #1 taking over a locally-managed property that was currently under public ownership, made the following assertion:

> For the good of the island, and if we need more room especially a family brand and really and truly as an island we need the family brand because we lost a lot of family rooms. So as an opportunity for a local person, it is a shame he would have to lose it….

There existed a perception that, if necessary, locals should give way to international ownership in the national interests of destination product development. This suggests a dependency relationship; TFDI was seen to have a direct impact on the country's ability to develop its tourism sector, a scenario suggested by Chaperon and Bramwell (2013). It may be seen to add weight to the importance of the ownership advantage possessed by TFDI properties as purported in Dunning's (1979) OLI framework.

Additionally, there was a positive relationship expressed between FDI and rapid tourism growth; since tourism growth is indicative of a multiplier in other sectors in the economy, this perceived correlation between FDI and rapid tourism growth insinuates a rapid growth in the supporting sectors as a result of FDI. In fact, this multiplier effect of tourism can be explicitly identified in the comments by respondents varying from "tourism is beneficial to local communities because it open doors for different services [like] taxi services" (Employee 3) to "… also people with little businesses can benefit from visitors" (Employee 4). This perceived increase in tourism correlated with TFDI also suggests that similar to those employed in these TFDI properties, TFDI has a substantial impact on generating employment in supporting services. The impact TFDI has on employment generation is an anticipated finding. However, what is perhaps more substantial and intriguing is the perception raised that TFDI creates an opportunity for individual local secondary service providers (in the value chain of the TFDI property) who have unique skills, to also gain access to opportunities in the wider value chain of TFDI-financed hotel chains (Employee 4).

Notwithstanding these positive benefits however, there was a recognition that the profit repatriation has potentially negative impacts on employees. As explained by one of the interviewees who works in a foreign-owned property:

> I think that too much foreign investment can hurt the community because not many of the benefits are contributed to the surrounding community since most the profits return to where the head company is based.

This recognition of leakages is generally accepted as a concern when discussing foreign ownership. However, it is particularly interesting that this TFDI employee made a direct association to the impact that these leakages could have on surrounding resident communities. The importance of this micro outlook of the impact of leakages by the respondent is reinforced by the fact that when reference is being made to employment and skills development opportunities, the respondents are first localising that impact in the community surrounding the foreign property.

Theme 3 – The Scale and Value of TFDI

Views on TFDI were mixed; employees from Hotel #1, Hotel #2, and Hotel #3 favoured TFDI growth. However, the interviews with employees of Hotel #5 and Hotel #6, in particular, highlighted concerns linked to the island's high dependency on tourism and the sustainability of tourism demand. The onus on local businesses was noted and there was a belief that there was potential to be more proactive in attracting tourists. The idea that FDI can have important positive effects on a host country's development effort (see for example, Davidson & Sahli, 2015; Samimi et al., 2013; Dinh et al., 2019) appears to be affected by the level of proactivity of local businesses. In this case, there appears to exist a recognition of potential benefits by foreign investors to be capitalised upon but questions abound around the seizing of those benefits. It was not simply a case of TFDI impact on tourism flows leading to net benefits to locals (Dwyer et al., 2003). These benefits needed to be secured through actions that focused on the sustainability of tourism demand. Otherwise, it was thought that there would not exist enough absolute demand to sustain local competition alongside international tourism entrants and there were concerns that country-specific location decisions of foreign investors might be negatively affected in the long run, namely with respect to any location advantages (Dunning, 1979) offered by Barbados. In line with the Economic Dependence Factor (Vargas-Sánchez et al., 2009), the local residents of Barbados, an island unanimously agreed upon as being heavily dependent on tourism, appeared to be favourable to tourism development.

Theme 4 – Power and Control

Power and control emerged as a key theme. This was evident in the views of employees from Hotel #2 when they acknowledged the role of the government in regulating FDI agreements. It also appeared in relation to a stated desire for mutually-beneficial tourism planning during Hotel #6 interviews and the consideration of the spread of benefits being restricted to government-level and hampered by economic leakages in the Hotel #5 interviews. Thus, attention was drawn to the relationships between foreign investors, the government, and local communities. The wider community-level impacts of TFDI were noted. There was a belief that foreign-owned hotels had inspired local businesses to improve their operations. One of the focus group interviewees, for example, proclaimed "for the longest while Hotel X [locally owned hotel] should have repaved their entrance but they would not, then Hotel Y [foreign-owned hotel] came in and all of a sudden things get improved". In terms of the relationship between TFDI entities and other local tourism providers, there was a sense amongst local residents that the former demonstrated a much more dynamic approach that was largely welcomed.

Community-based initiatives such as Hotel #1's Foundation were acknowledged as genuine attempts by the foreign investors to create positive local impacts. There was a perceived difference in the ways in which international firms responded to environmental issues compared to local firms. However, there was a sense of caution about the performance of foreign-owned investors amongst some of the residents interviewed due to failure of some of the hotels to deliver on "promises". The official measurement of the positive impacts offered by these investors was questioned and it was suggested that economic leakages may in fact be underestimated due, in part, to a government focus on high-level tourist volume and value figures rather than paying attention to the experiences presented at a micro or "ground level".

Conclusion

While a lot has been written about FDI impacts on a country's economic development, it is fair to say that there has been insufficient empirical evidence linking FDI and local residents' perceptions. This chapter has contributed to knowledge creation in terms of TFDI impact in an under-researched context, Barbados and SIDS. It has presented a critical evaluation of the "realities" of TFDI as experienced by local hotel employees and identified relationships between these perceptions and currently known impacts of tourism FDI. It may be argued that the perceptions gathered during this research study are not merely a reflection on the self-interests of local hotel employees. For instance, observations of the power dynamics between groups, namely, foreign-owned investors and the government was a theme that was highlighted.

An area for further exploration is the SIDS context that formed a background for the case study setting in this chapter. Further qualitative research is needed on the local impacts of tourism FDI in SIDS. To date, tourism researchers have tended to focus their analyses on restricted themes (see for example Pratt, 2015) and/or limited stakeholder views (see, for example, Prayag et al., 2010). It is recommended that a more holistic examination of economic, socio-cultural, and environmental impacts is undertaken that accommodates the voices and experiences of multiple stakeholders. It is notable that where research exists, it lacks a qualitative focus. A richer, in-depth analysis is important to help strategic players in their ability to further enhance human and tourism development, by minimising the negative and increasing the positive impacts of TFDI.

The aim of this chapter is to offer valuable insights for better understanding the perceived impacts of tourism FDI on local hotel employees in a tourism-dependent SIDS, and more widely in other SIDS where local realities have often gone unnoticed.

References

Alam, M.S., & Paramati, S.R. (2016). The impact of tourism on income inequality in developing economies: Does Kuznets curve hypothesis exist? *Annals of Tourism Research*, *61*, pp. 111–126.

Almeida-García, F., Peláez-Fernández, M.Á., Balbuena-Vázquez, A., & Cortés-Macias, R. (2016). Residents' perceptions of tourism development in Benalmádena (Spain). *Tourism Management*, *54*, pp. 259–274.

Arain, H., Han, L., Sharif, A., & Meo, M.S. (2020). Investigating the effect of inbound tourism on FDI: The importance of quantile estimations. *Tourism Economics*, *26*(4), pp. 682–703.

Barbados Ministry of Finance and Economic Affairs (2011). Barbados economic and social report 2010, *Research and Planning Unit Ministry of Finance and Economic Affairs*. Retrieved from http://www.economicaffairs.gov.bb/archive-detail.php?id=277

Barbados Tourism Investment Inc. (BTI) (2015). *Tourism Profile*. Retrieved from https://www.barbadostourisminvestment.com/tourism-profile/

Barbados Tourism Master Plan (2014). *Barbados Tourism Master Plan 2014–2023*. Retrieved from http://www.bhta.org/images/TMP/TMP-Report-I.pdf

Bayulgen, O. (2010) *Foreign Investment and Political Regimes: The Oil Sector in Azerbaijan, Russia and Norway*. Cambridge: Cambridge University Press.

Bianchi, R. (2018). The political economy of tourism development: A critical review. *Annals of Tourism Research*, *70*, pp. 88–102.

Bradshaw, M. (2015). Sandals sparks beach worry. *Nation Barbados*. Retrieved from http://www.nationnews.com/nationnews/news/62250/sandals-sparks-beach-worry

Brohman, J. (1996). New directions for tourism in the Third World, *Annals of Tourism Research, 9*(3), pp. 331–358

Buckley, P., & Casson, P. (1976). *The Future of the Multinational Enterprise.* NY: Holmes & Meier.

Bury, J. (2001). Corporations and capitals: A framework for evaluating the impacts of transnational corporations in developing countries. *Journal of Corporate Citizenship, 1*, pp. 75–91.

Central Bank of Barbados (2010). *The Evolution of the Foreign- Exchange Earning Sectors in Barbados. A Presentation for the 2010 National Consultation.* Retrieved May 21, 2012 http://www.centralbank.org.bb/WEBCBB.nsf/vwNews/CD55FAA603196702042576EF004FDA36/$FILE/Natl_Consulation-Draft1.1.pdf.

Chaperon, S., & Bramwell, B. (2013). Dependency and agency in peripheral tourism development. *Annals of Tourism Research, 40*, pp. 132–154.

Chirenje, L.I., Chitotombe, J., Gukurume, S., Chazovachii, B. and Chitongo, L. (2013). The impact of tourism leakages on local economies: a case study of Nyanga District, Zimbabwe. *Journal of Human Ecology, 42*(1), pp. 9–16.

Coles, T., Fenclova, E., & Dinan, C. (2013). Tourism and corporate social responsibility: A critical review and research agenda. *Tourism Management Perspectives, 6*, pp. 122–141.

Davidson, L., & Sahli, M. (2015). Foreign direct investment in tourism, poverty alleviation, and sustainable development: A review of the Gambian hotel sector. *Journal of Sustainable Tourism, 23*(2), pp. 167–187.

Dinh, T.T.H., Vo, D.H., & Nguyen, T.C. (2019). Foreign direct investment and economic growth in the short run and long run: Empirical evidence from developing countries. *Journal of Risk and Financial Management, 12*(4), p.176.

Dunning, J.H. (1979). Explaining changing patterns of international production: In defense of the eclectic theory. *Oxford Bulletin of Economics and Statistics, 41*, pp. 269–295.

Dunning, J.H. (2001). The Eclectic (OLI) Paradigm of International Production: Past, Present and Future. *International Journal of the Economics of Business, 18*(2), pp. 173–190.

Dwyer, L., Forsyth, P., & Spurr, R. (2003). Inter-industry effects of tourism growth: Implications for destination managers. *Tourism Economics, 9*(2), pp. 117–132.

Fauzel, S. (2016). Modeling the relationship between FDI and financial development in small island economies: A PVAR approach. *Theoretical Economics Letters, 6*, pp. 367–375.

Fauzel, S. (2020). FDI and tourism futures: A dynamic investigation for a panel of small island economies. *Journal of Tourism Futures, 7*(1), pp. 98–110.

Gupta, R. (2015). Tourism, FDI and inclusive growth in India. *Pacific Business Review International, 7*(12), pp. 108–114.

Hymer, S. (1976) (1960 dissertation). *The International Operations of Nation Firms: A Study of Foreign Direct Investment*, Cambridge: MLT Press.

Jönsson, C. (2005). Barbados Casuarina Beach Resort – A model for the use of cleaner technologies in the hotel sector?. *Proceedings 2nd Tourism Outlook Conference – Tourism Edge and Beyond, Selangor: Malaysia*, pp. 223–237.

Jönsson, C. (2008). Overcoming obstacles of accommodation shortage in preparation for the ICC cricket world cup 2007: Community attitude to Barbados' alternative accommodation initiative. *London Journal of Tourism, Sport and Creative Industries, 1*(2), Autumn, pp. 21–38.

Jönsson, C., & Harrison, N. (2015). Targeting a new tourism market: Is Barbados ready for a new market? In Ray, N., Kumar Das, D., Ghosh, A., & Chaudhuri, S. (Eds.) *Strategic Infrastructure Development for Economic Growth and Social Change.* , USA: IGI Global Publication.

Kaulihowa, T. and Adjasi, C., 2018. FDI and income inequality in Africa. *Oxford Development Studies, 46*(2), pp. 250–265.

Khan, A., Bibi, S., Lorenzo, A., Lyu, J., & Babar, Z.U. (2020). Tourism and development in developing economies: A policy implication perspective. *Sustainability, 12*(4), p.1618.

Meyer, D. (2006). Foreign direct investment in tourism: The development dimension. Case Study: Sri Lanka, *Report for UNCTAD Project.*

Meyer, D. (2013). Exploring the duality of structure and agency–the changing dependency paradigms of tourism development on the Swahili coast of Kenya and Zanzibar. *Current Issues in Tourism, 16*(7–8), pp. 773–791.

Mustafa, A.M.M. (2019). Contribution of tourism and foreign direct investment to gross domestic product: Econometric analysis in the case of Sri Lanka. *The Journal of Asian Finance, Economics and Business, 6*(4), pp. 109–114.

Nunkoo, R., & Ramkissoon, H. (2011). Developing a community support model for tourism. *Annals of Tourism Research, 38*(3), 964–988.

Pratt, S. (2015). The economic impact of tourism in SIDS. *Annals of Tourism Research, 52*, pp. 148–160.

Prayag, G., Dookhony-Ramphul, K., & Maryeven, M. (2010). Hotel development and tourism impacts in Mauritius: Hoteliers' perspectives on sustainable tourism. *Development Southern Africa, 27*(5), 697–712.

Samimi, A.J., Sadeghi, S., & Sadeghi, S. (2013). The relationship between foreign direct investment and tourism development: Evidence from developing countries. *Institutions and Economies, 5*(2), pp. 59–68.

Sapir, E. (1958). *Culture, Language and Personality.* Mandelbaum, D.G. (Ed.). Berkeley, CA: University of California Press.

Schulte, D.C., & Dissart, J. (2015). Resorts, hinterlands and local developments. In Dissart, J., Dehez, J., & Marsat, J. (Eds.), *Tourism, Recreation and Regional Development: Perspectives from France and Abroad.* Surrey: Ashgate Publishing Ltd. pp. 155–74.

Sharpley, R. (2014). Host perceptions of tourism: A review of the research. *Tourism Management, 42*, pp. 37–49.

Simão, J. & Môsso, A. (2013). Residents' perceptions towards tourism development: The case of Sal Island. *International Journal of Development Issues, 12*(2), pp. 140–157.

Trends in FDI Chapter 1.

UNCTAD (2007). *World Investment Report 2007.* United Nations Conference on Trade and Development, Geneva.

UNCTAD (2010). *World Investment Report 2010: Investing in a Low-Carbon Economy,* Global Trends in FDI Chapter 1. Geneva: United Nations.

UNCTAD (2016). *A Commitment to Inclusive Trade.* Geneva: United Nations.

UNCTAD (2020). *World Investment Report 2020: International Production beyond the Pandemic.* Geneva: United Nations.

UNWTO (2021a). *Tourist Numbers Down by 83% but Confidence Slowly Rising*, June 2. Retrieved: https://www.unwto.org/taxonomy/term/347

UNWTO (2021b). *World Tourism Organization, Yearbook of Tourism Statistics, Compendium of Tourism Statistics and Data Files International Tourism, Barbados.* Retrieved: https://data.worldbank.org/indicator/ST.INT.ARVL?locations=BB

United Nations World Tourism Organization (UNWTO). (2017). *Tourism Highlights 2017 Edition*, Retrieved from http://www.e-unwto.org/doi/pdf/10.18111/9789284419029

Vargas-Sánchez, A., Plaza-Mejía, M., de los Á., & Porras-Bueno, N. (2009). Understanding residents' attitudes toward the development of industrial tourism in a former mining community. *Journal of Travel Research*, *47*(3), pp. 373–387.

Viscusi, K., Harrington, J., & Vernon, J. (2005). *Economics of Regulation and Antitrust.* Massachusetts: Massachusetts Institute of Technology

Vítová, P., Harmáček, J., & Opršal, Z. (2019). Determinants of tourism flows in small island developing states (SIDS). *Island Studies Journal*, *14*(2), pp. 3–22.

World Bank (2015). *Ease of Doing Business Report 2015- Going Beyond Efficiency.* Retrieved from http://www.doingbusiness.org/content/dam/doingBusiness/media/Annual-Reports/English/DB15-Full-Report.pdf

World Bank (2017). *Global Investment Competitiveness Report 2017/2018: Foreign Investor Perspectives and Policy Implications.*

World Bank (2018). *Barbados Trade Summary 2016.* Retrieved from https://wits.worldbank.org/CountryProfile/en/Country/BRB/Year/LTST/Summarytext

Wreath, R. (2015). *Sandals Barbados – Elegance, Luxury and Glamour.* Retrieved from http://www.telegraph.co.uk/sponsored/travel/beautiful-barbados/11586571/barbados-holidays-elegance-luxury-glamour.html

WTTC, M. (2002). *The Impact of Travel & Tourism on Jobs and the Economy.* Queen Victoria: World Travel & Tourism Council.

WTTC (2017). *Economic Impact of Travel & Tourism (Barbados) 2017.* Annual Update: Summary. Retrieved from https://www.wttc.org/-/media/files/reports/economic-impact-research/archived/countries-2017-old/barbados2017.pdf

Part IV
Conclusions

13 Current Knowledge and Future Research Directions in Tourism Foreign Direct Investment

H. Cristina Jönsson

Introduction

It is the aim of this book to further our understanding of TFDI using current empirical research and case studies. Chapters contributed by international tourism scholars discuss theoretical and practical approaches towards TFDI, as well as its impacts and role in different geographical settings. This final chapter summarises the issues and aspects highlighted, as well as conclusions formulated by authors in previous chapters. Suggestions and recommendations for local planners, destination and business managers to successfully manage tourism FDI are also provided in this chapter. The key findings from this book are reviewed in order to derive a conclusion as well as future research directions.

This book covers a wide but by no means exhaustive range of TFDI impacts. All chapters share the assumption that while economic theory predicts a positive impact of FDI on economic growth, empirical evidence is mixed.

Sustainable Tourism Foreign Direct Investment in Context

Tourism FDI has emerged as an important factor in achieving sustainable economic growth for both developed and developing countries. Despite this, there have been years of arguments in favour of and against FDI, starting with Dunning and Kundu (1982). As indicated in Chapter 1, TFDI has become a key component of national development strategies for many countries and the growth of TFDI has been phenomenal. Chapter 2 conceptually examines potential advantages and disadvantages of FDI in tourism. It is argues that governments reserve considerable power to determine how the tourism sector performs while policies can be endorsed to increase local economic participation. This can lead to reduced economic dependencies on other countries. As TFDI is linked to both positive and negative effects on the tourism industry of a host destination, the author cautions against claiming TFDI as a potential catalyst for tourism industry and economic development. Importantly, this chapter points to the careful

DOI: 4324/9781003155492-17

examination of policies concerning FDI and MNE involvement in tourism, within the framework of overall development strategies, weighing up the socio-economic benefits that result against the possible associated costs to destination sustainable development

Chapter 3 follows-up on the theme of development strategies and impacts by bringing together ideas that may be useful in developing an overarching framework or lens to examine local perceptions of TFDI in a case study location. This offers an opportunity to shift away from traditional neo-liberalist theories in the study of tourism and FDI. It also suggests consideration of stewardship of more inclusive economic and social development. Notably, the discussion presented in this chapter provides a useful approach to better understand the impacts of TFDI on local people in a tourism-dependent, small island economy and potentially in other tourism economies where local resident voices have often gone unnoticed.

Based on the VECM framework, Chapter 4 investigates the relationship that might exist between TFDI and sustainable development in Mauritius. It applies a recent co-integration approach in a dynamic framework and presents new evidence in the field of TFDI and sustainable development. The author highlights a significant amount of TFDI that have contributed towards the sustainable development of Mauritius and concludes that to further boost sustainable development in Mauritius, there are various actions that can still be taken in the areas of, for example, governance and eco-conscious policies.

Part II ends with Chapter 5, which highlights the value of exploring an evolutionary approach to investigating TFDI in the context of a transitional, post-Soviet economy through an evolutionary lens. It demonstrates the scope to gain insights into both host government investment economy and foreign investor behaviours and actions over a sustained time-period. The author maintains that the approach in this chapter has assisted identification of issues pertinent to the lasting interest of a TFDI in Kyrgyzstan and it has provided a fuller understanding of tourism processes underpinning development. The approach has facilitated consideration of both structure and agency in tourism investment. Limitations of relying solely on secondary data analysis are highlighted and the author indicates the potential for further research to gain a deeper understanding of how and why "specific paths unfold in particular ways over time and space" (Pike et al., 2016:131).

Overall, this set of introductory chapters raises three important issues that are to date underrepresented in the relevant literature; first, the relationship between FDI and tourism is addressed both conceptually and practically; second, theories and frameworks related to tourism FDI are explored. It is important to note that both Chapters 4 and 5 suggest that it is necessary to attract sustainable FDI. Third, Chapter 3 in this section is one of the few works that draws attention to the importance of local voices for a greater understanding of TFDI impacts and its role in tourism development.

Current Issues in Tourism Foreign Direct Investment

Determinants of French inward FDI stock in the hospitality industry are the topic of Chapter 6. A panel gravity model is applied to bilateral inward FDI stock between France and 19 investor countries in the Hotels and Restaurants industry. While the cost-based considerations such as labour costs and taxes are an important determinant factor of hospitality FDI activity, overall, the authors find that the most significant determinants of FDI in the hospitality industry are practically the same as those obtained for total FDI stock. The study results in this chapter have two important implications for tourism policymakers. Firstly, the control of cost factors such as labour costs and taxes is crucial for countries to remain competitive. Secondly, adverse effects of high taxation can be mitigated by the provision of public goods.

The analysis of the impacts of FDI on Croatian tourism is the topic of Chapter 7. The poor structure of accommodation capacities in relation to other EU member states is clearly a problem. The author therefore suggests that there is a need for commitment to attract FDI in tourism, with special emphasis on global hotel brands.

Chapter 8 examines the causal link between FDI and tourism in Sub-Saharan African (SSA) countries by studying the short and long term impacts of FDI growth on tourism development using the autoregressive distributed lag model. The data analysis indicates that FDI and economic growth are key drivers for tourism growth in the long run for SSA countries. The author therefore calls for the implementation of FDI promotion programs that focus on tourism development and economic growth.

There are few studies that analyse the relationship between FDI and tourism growth. One of the reasons for this is the lack of FDI data at the sectorial level. The authors of Chapter 9 aim to fill that gap by analysing the relation between FDI and tourism growth in China by measuring international tourist arrivals and international tourism receipts. The result presented in this chapter provides insights into the role FDI plays in stimulating tourism growth, and how tourism can be used as a tool to attract FDI. The authors suggest that policy makers promote sustainable tourism development by paying attention to tourism development policies that explicitly promote sustainable tourism development and low carbon emissions.

The focus of Chapter 10 is nature tourism FDI in Bangladesh. The chapter study arrives at a model that explains and predicts how economic development can be viable through nature tourism growth. The author considers that the model, if implemented as developed, will be a guide to an attractive destination both for tourists and FDI, and may lead to accelerated economic development.

Chapter 11 examines the effect of TFDI on economic growth in Indonesia, focusing on human capital, trade openness and financial market development. The relationship between TFDI and economic growth is under researched in Indonesia. The author therefore aims to fill this research gap.

Based on the results of the chapter study, several recommendations are made both for further research, as well as for practitioners.

The consideration of local residents' perceptions of tourism FDI impacts in Chapter 12 gives rise to an approach to FDI impact studies that goes beyond the widely used econometric and statistical studies of tourism FDI. Here, the "realities" of local hotel employees are considered as an important factor in determining tourism FDI impacts. This chapter is one of the few works that focuses on TFDI in Barbados and the Caribbean. The author suggests that a more holistic examination of economic, socio-cultural and environmental impacts of TFDI is undertaken that accommodates the voices and experiences of multiple stakeholders.

Conclusions

The evidence examined in this book offers a basis for ample commentary on the issues, challenges and prospects associated with TFDI. The details of the research and major arguments presented are too complex to be straightforwardly summarised. However, evidence gathered in this book demonstrates that a search for a universal result of TFDI impacts is simply misguided. TFDI can have dramatically differing impacts, both positive and negative; therefore, it is imperative to undertake more holistic research of TFDI, by including the voices, perceptions and experiences of multiple TFDI stakeholders.

This book provides insights into theoretical as well as practical issues by covering conceptual foundations, critical concepts and current issues. It addresses numerous gaps in our understanding of FDI in tourism destinations. That said, despite its wide range of topics, themes and case studies, no book can completely cover TFDI. If anything, the empirical studies presented here shed light on an even wider range of possible research topics that undoubtedly would further enhance our knowledge of TFDI.

There are future challenges for TFDI that could not be covered in this book. These are mainly due to COVID-19 and the global pandemic, which has led to a decrease in tourism demand. The issues and the possibilities the pandemic offers tourism destinations must be contemplated in order to understand future tourism development and FDI. The future challenge of using TFDI as part of development in tourism is much more complicated than conventional wisdom suggests. To capture the full advantages of TFDI it requires a much broader and more energetic action agenda than developing countries and tourism economies have been accustomed to pursue. By considering these and future issues in tourism development and FDI, this book will hopefully prove informative and inspirational for tourism development decision makers and practitioners, as well as researchers.

References

Dunning, J. and Kundu, S.K., 1982. *Impact of Transnational Hotel Chains and Management Companies Operating in Developing Countries u Transnational Corporations in International Tourism, United Nations.* New York and Geneva: United Nations.

Pike, A., MacKinnon, D., Cumbers, A., Dawley, S. and McMaster, R., 2016. Doing evolution in economic geography. *Economic Geography, 92*(2), pp. 123–144.

Index

Note: **Bold** page numbers refer to tables and *italic* page numbers refer to figures.

Taylor & Francis Group
an **informa** business

Taylor & Francis eBooks

www.taylorfrancis.com

A single destination for eBooks from Taylor & Francis with increased functionality and an improved user experience to meet the needs of our customers.

90,000+ eBooks of award-winning academic content in Humanities, Social Science, Science, Technology, Engineering, and Medical written by a global network of editors and authors.

TAYLOR & FRANCIS EBOOKS OFFERS:

A streamlined experience for our library customers

A single point of discovery for all of our eBook content

Improved search and discovery of content at both book and chapter level

REQUEST A FREE TRIAL

support@taylorfrancis.com

Routledge
Taylor & Francis Group

CRC Press
Taylor & Francis Group